국제환경조약집

박 덕 영

세창출판사

이 도서의 국립중앙도서관 출판시도서목록(CIP)은 e-CIP 홈페이지 (http://www.nl.go.kr/ecip)에서 이용하실 수 있습니다.(CIP제어번호: CIP2010002549)

머 리 말

오랫동안 국제환경법, 특히 무역과 환경문제에 관심을 두고 있었다. 최근 기후변화 문제가 국제사회의 주요 관심사가 되었고, 국내적으로는 이명박 정부 들어 지난 몇 년간 녹색성장이 커다란 화두가 되고 있다. 환경 문제를 이해하는 데 있어서는 국제조약의 이해가 반드시 필요하고, 우리나라처럼 수출이 국가 성장의 동력이 되는 나라에서는 국제환경법과 국제환경조약, 각 국가들의 환경규제 문제를 이해하는 것이 필수적인 요소가 되고 있다.

그럼에도 불구하고 아직도 국내에 종합적인 국제환경조약집 하나 없다는 아쉬움을 품고 있던 차에 조약집을 편집하기로 마음먹었다. 국제환경조약에 관한 방대한 자료를 수집하고 정리하였으나, 국내 출판시장의 어려움으로 인하여 우선 기후변화와 생물다양성에 관한 조약들만을 모아 자그마한 영문 조약집으로 출간하게 되었다. 앞으로 해양오염 문제, 위험물질의 이동, 환경책임과 절차적인 문제에 관한 조약들을 모아 추후 작업을 할 예정이다.

환경문제의 중요성이 증가함에 따라 국내 각 대학과 로스쿨에서 국제환경법이 강의될 예정이다. 과거에 없던 강의가 새로이 개설되고 있고, 환경법 강의든 국제환경법 강의든 환경 문제를 이해하는 데 있어서 국제조약에 대한 이해는 반드시 필요한 부분이다. 본서가 이에 조금이나마 보탬이 되기를 기대한다.

어려운 여건에도 출판을 허락하여 주신 세창출판사 이방원 사장님과 임길남 상무님, 본서를 깔끔하게 편집해 주신 안효희님께 감사드린다. 그리고 본서의 출판을 주선해 주신 산하온 환경연구소장 김기순 박사님께도 감사의 말씀을 올린다.

마지막으로 편집과정을 도와준 연세대학교 대학원 석사과정 조유미 양에게 고마움을 전하며, 앞으로 학문적 성공이 있기를 기원한다.

2010년 여름

박 덕 영 씀

차 례

Part I. General

Part II. Climate Change

Part Ⅲ. Biodiversity

Part I. General

1. Stockholm Declaration on the Human Environment (1972)

Date: 16 June 1972
Link: www.unep.org

The United Nations Conference on the Human Environment,

Having met at Stockholm from 5 to 16 June 1972,

Having considered the need for a common outlook and for common principles to inspire and guide the peoples of the world in the preservation and enhancement of the human environment,

Part I

Proclaims that:

1. Man is both creature and moulder of his environment, which gives him physical sustenance and affords him the opportunity for intellectual, moral, social and spiritual growth. In the long and tortuous evolution of the human race on this planet a stage has been reached when, through the rapid acceleration of science and technology, man has acquired the power to transform his environment in countless ways and on an unprecedented scale. Both aspects of man's environment, the natural and the man-made, are essential to his well-being and to the enjoyment of basic human rights even the right to life itself.

2. The protection and improvement of the human environment is a major issue which affects the well-being of peoples and economic development throughout the world; it is the urgent desire of the peoples of the whole world and the duty of all Governments.

3. Man has constantly to sum up experience and go on discovering, inventing, creating and advancing. In our time, man's capability to transform his surroundings, if used wisely, can bring to all peoples the benefits of development and the opportunity to enhance the quality of life. Wrongly or heedlessly applied, the same power can do incalculable harm to human beings and the human environment. We see around us growing evidence of man-made harm in many regions of the earth: dangerous levels of pollution in water, air, earth and living beings; major and undesirable disturbances to the ecological balance of the biosphere; destruction and depletion of irreplaceable resources; and gross deficiencies, harmful to the physical, mental and social health of man, in the man-made environment, particularly in the living and working environment.

4. In the developing countries most of the environmental problems are caused by under-development. Millions continue to live far below the minimum levels required for a decent human existence, deprived of adequate food and clothing, shelter and education, health and sanitation. Therefore, the developing countries must direct their efforts to development, bearing in mind their priorities and the need to safeguard and improve the environment. For the same purpose, the industrialized countries should make efforts to reduce the gap themselves and the developing countries. In the industrialized countries, environmental problems are generally related to industrialization and technological development.

5. The natural growth of population continuously presents problems for the preservation of the environment, and adequate policies and measures should be adopted, as appropriate, to face these problems. Of all things in the world, people are the most precious. It is the people that propel social progress, create social wealth, develop science and technology and, through their hard work, continuously transform the human environment. Along with

social progress and the advance of production, science and technology, the capability of man to improve the environment increases with each passing day.

6. A point has been reached in history when we must shape our actions throughout the world with a more prudent care for their environmental consequences. Through ignorance or indifference we can do massive and irreversible harm to the earthly environment on which our life and well-being depend. Conversely, through fuller knowledge and wiser action, we can achieve for ourselves and our posterity a better life in an environment more in keeping with human needs and hopes. There are broad vistas for the enhancement of environmental quality and the creation of a good life. What is needed is an enthusiastic but calm state of mind and intense but orderly work. For the purpose of attaining freedom in the world of nature, man must use knowledge to build, in collaboration with nature, a better environment. To defend and improve the human environment for present and future generations has become an imperative goal for mankind a goal to be pursued together with, and in harmony with, the established and fundamental goals of peace and of worldwide economic and social development.

7. To achieve this environmental goal will demand the acceptance of responsibility by citizens and communities and by enterprises and institutions at every level; all sharing equitably in common efforts. Individuals in all walks of life as well as organizations in many fields, by their values and the sum of their actions, will shape the world environment of the future. Local and national governments will bear the greatest burden for large-scale environmental policy and action within their jurisdictions. International co-operation is also needed in order to raise resources to support the developing countries in carrying out their responsibilities in this field. A growing class of environmental problems, because they are regional or global in extent or because

they affect the common international realm, will require extensive co-operation among nations and action by international organizations in the common interest. The Conference calls upon Governments and peoples to exert common efforts for the preservation and improvement of the human environment, for the benefit of all the people and for their posterity.

Part II

Principles

States the common conviction that:

Principle 1

Man has the fundamental right to freedom, equality and adequate conditions of life, in an environment of a quality that permits a life of dignity and well-being, and he bears a solemn responsibility to protect and improve the environment for present and future generations. In this respect, policies promoting or perpetuating apartheid, racial segregation, discrimination, colonial and other forms of oppression and foreign domination stand condemned and must be eliminated.

Principle 2

The natural resources of the earth, including the air, water, land, flora and fauna and especially representative samples of natural ecosystems, must be safeguarded for the benefit of present and future generations through careful planning or management, as appropriate.

Principle 3

The capacity of the earth to produce vital renewable resources must be maintained and, wherever practicable, restored or improved.

Principle 4

Man has a special responsibility to safeguard and wisely manage the heritage of wildlife and its habitat, which are now gravely imperilled by a combination of adverse factors. Nature conservation, including wildlife, must therefore receive importance in planning for economic development.

Principle 5

The non-renewable resources of the earth must be employed in such a way as to guard against the danger of their future exhaustion and to ensure that benefits from such employment are shared by all mankind.

Principle 6

The discharge of toxic substances or of other substances and the release of heat, in such quantities or concentrations as to exceed the capacity of the environment to render them harmless, must be halted in order to ensure that serious or irreversible damage is not inflicted upon ecosystems. The just struggle of the peoples of all countries against pollution should be supported.

Principle 7

States shall take all possible steps to prevent pollution of the seas by substances that are liable to create hazards to human health, to harm living resources and marine life, to damage amenities or to interfere with other legitimate uses of the sea.

Principle 8

Economic and social development is essential for ensuring a favourable living and working environment for man and for creating conditions on earth that are necessary for the improvement of the quality of life.

Principle 9

Environmental deficiencies generated by the conditions of under-development and natural disasters pose grave problems and can best be remedied by accelerated development through the transfer of substantial quantities of financial and technological assistance as a supplement to the domestic effort of the developing countries and such timely assistance as may be required.

Principle 10

For the developing countries, stability of prices and adequate earnings for primary commodities and raw materials are essential to environmental management since economic factors as well as ecological processes must be taken into account.

Principle 11

The environmental policies of all States should enhance and not adversely affect the present or future development potential of developing countries, nor should they hamper the attainment of better living conditions for all, and appropriate steps should be taken by States and international organizations with a view to reaching agreement on meeting the possible national and international economic consequences resulting from the application of environmental measures.

Principle 12

Resources should be made available to preserve and improve the environment, taking into account the circumstances and particular requirements of developing countries and any costs which may emanate from their incorporating environmental safeguards into their development planning and the need for making available to them, upon their request, additional international technical and financial assistance for this purpose.

Principle 13

In order to achieve a more rational management of resources and thus to improve the environment, States should adopt an integrated and co-ordinated approach to their development planning so as to ensure that development is compatible with the need to protect and improve environment for the benefit of their population.

Principle 14

Rational planning constitutes an essential tool for reconciling any conflict between the needs of development and the need to protect and improve the environment.

Principle 15

Planning must be applied to human settlements and urbanization with a view to avoiding adverse effects on the environment and obtaining maximum social, economic and environmental benefits for all. In this respect, projects which are designed for colonialist and racist domination must be abandoned.

Principle 16

Demographic policies which are without prejudice to basic human rights and which are deemed appropriate by Governments concerned should be applied in those regions where the rate of population growth or excessive population concentrations are likely to have adverse effects on the environment of the human environment and impede development.

Principle 17

Appropriate national institutions must be entrusted with the task of planning, managing or controlling the environmental resources of States with a view to enhancing environmental quality.

Principle 18

Science and technology, as part of their contribution to economic and social development, must be applied to the identification, avoidance and control of environmental risks and the solution of environmental problems and for the common good of mankind.

Principle 19

Education in environmental matters, for the younger generation as well as adults, giving due consideration to the underprivileged, is essential in order to broaden the basis for an enlightened opinion and responsible conduct by individuals, enterprises and communities in protecting and improving the environment in its full human dimension. It is also essential that mass media of communications avoid contributing to the deterioration of the environment, but, on the contrary, disseminate information of an educational nature on the need to protect and improve the environment in order to enable man to develop in every respect.

Principle 20

Scientific research and development in the context of environmental problems, both national and multinational, must be promoted in all countries, especially the developing countries. In this connection, the free flow of up-to-date scientific information and transfer of experience must be supported and assisted, to facilitate the solution of environmental problems; environmental technologies should be made available to developing countries on terms which would encourage their wide dissemination without constituting an economic burden on the developing countries.

Principle 21

States have, in accordance with the Charter of the United Nations and the principles of international law, the sovereign right to exploit their own resources pursuant to their own environmental policies, and the responsibility to ensure that activities within their jurisdiction or control do not cause damage to the environment of other States or of areas beyond the limits of national jurisdiction.

Principle 22

States shall co-operate to develop further the international law regarding liability and compensation for the victims of pollution and other environmental damage caused by activities within the jurisdiction or control of such States to areas beyond their jurisdiction.

Principle 23

Without prejudice to such criteria as may be agreed upon by the international community, or to standards which will have to be determined nationally, it will be essential in all cases to consider the systems of values prevailing in each country, and the extent of the applicability of standards which are valid for the most advanced countries but which may be inappropriate and of unwarranted social cost for the developing countries.

Principle 24

International matters concerning the protection and improvement of the environment should be handled in a co-operative spirit by all countries, big and small, on an equal footing. Co-operation through multilateral or bilateral arrangements or other appropriate means is essential to effectively control, prevent, reduce and eliminate adverse environmental effects resulting from activities conducted in all spheres, in such a way that due account is taken of the sovereignty and interests of all States.

Principle 25

States shall ensure that international organizations play a co-ordinated, efficient and dynamic role for the protection and improvement of the environment.

Principle 26

Man and his environment must be spared the effects of nuclear weapons and all other means of mass destruction. States must strive to reach prompt agreement, in the relevant international organs, on the elimination and complete destruction of such weapons.

2. UN GA Resolution 37/7 on a World Charter for Nature (1982)

Date: 28 October 1982
Link: www.un.org

The General Assembly,

Having considered the report of the Secretary-General on the revised draft World Charter for Nature,

Recalling that, in its resolution 35/7 of 30 October 1980, it expressed its conviction that the benefits which could be obtained from nature depended on the maintenance of natural processes and on the diversity of life forms and that those benefits were jeopardized by the excessive exploitation and the destruction of natural habitats,

Further recalling that, in the same resolution, it recognized the need for appropriate measures at the national and international levels to protect nature and promote international co-operation in that field,

Recalling that, in its resolution 36/6 of 27 October 1981, it again expressed its awareness of the crucial importance attached by the international community to the promotion and development of co-operation aimed at protecting and safeguarding the balance and quality of nature and invited the Secretary-General to transmit to Member States the text of the revised version of the draft World Charter for Nature contained in the report of the Ad Hoc Group of Experts on the draft World Charter for Nature, as well as any further observations by States, with a view to appropriate consideration by the General Assembly at its thirty-seventh session,

Conscious of the spirit and terms of its resolutions 35/7 and 36/6, in which it solemnly invited Member States, in the exercise of their permanent sovereignty over their natural resources, to conduct their activities in recognition of the supreme importance of protecting natural systems, maintaining the balance and quality of nature and conserving natural resources, in the interests of present and future generations,

Having considered the supplementary report of the Secretary-General,

Expressing its gratitude to the Ad Hoc Group of Experts which, through its work, has assembled the necessary elements for the General Assembly to be able to complete the consideration of and adopt the revised draft World Charter for Nature at its thirty-seventh session, as it had previously recommended,

Adopts and solemnly proclaims the World Charter for Nature contained in the annex to the present resolution.

ANNEX
World Charter for Nature

The General Assembly,

Reaffirming the fundamental purposes of the United Nations, in particular the maintenance of international peace and security, the development of friendly relations among nations and the achievement of international co-operation in solving international problems of an economic, social, cultural, technical, intellectual or humanitarian character,

Aware that:

(a) Mankind is a part of nature and life depends on the uninterrupted functioning of natural systems which ensure the supply of energy and nutrients,

(b) Civilization is rooted in nature, which has shaped human culture and influenced all artistic and scientific achievement, and living in harmony with nature gives man the best opportunities for the development of his creativity, and for rest and recreation,

Convinced that:

(a) Every form of life is unique, warranting respect regardless of its worth to man, and, to accord other organisms such recognition, man must be guided by a moral code of action,

(b) Man can alter nature and exhaust natural resources by his action or its consequences and, therefore, must fully recognize the urgency of maintaining the stability and quality of nature and of conserving natural resources,

Persuaded that:

(a) Lasting benefits from nature depend upon the maintenance of essential ecological processes and life support systems, and upon the diversity of life forms, which are jeopardized through excessive exploitation and habitat destruction by man,

(b) The degradation of natural systems owing to excessive consumption and misuse of natural resources, as well as to failure to establish an appropriate economic order among peoples and among States, leads to the breakdown of the economic, social and political framework of civilization,

(c) Competition for scarce resources creates conflicts, whereas the conservation of nature and natural resources contributes to justice and the maintenance of peace and cannot be achieved until mankind learns to live in peace and to forsake war and armaments,

Reaffirming that man must acquire the knowledge to maintain and enhance his ability to use natural resources in a manner which ensures the preservation of the species and ecosystems for the benefit of present and future generations,

Firmly convinced of the need for appropriate measures, at the national and international, individual and collective, and private and public levels, to protect nature and promote international co-operation in this field,

Adopts, to these ends, the present World Charter for Nature, which proclaims the following principles of conservation by which all human conduct affecting nature is to be guided and judged.

I. GENERAL PRINCIPLES

1. Nature shall be respected and its essential processes shall not be impaired.

2. The genetic viability on the earth shall not be compromised; the population levels of all life forms, wild and domesticated, must be at least sufficient for their survival, and to this end necessary habitats shall be safeguarded.

3. All areas of the earth, both land and sea, shall be subject to these principles of conservation; special protection shall be given to unique areas, to representative samples of all the different types of ecosystems and to the habitats of rare or endangered species.

4. Ecosystems and organisms, as well as the land, marine and atmospheric resources that are utilized by man, shall be managed to achieve and maintain optimum sustainable productivity, but not in such a way as to endanger the integrity of those other ecosystems or species with which they coexist.

5. Nature shall be secured against degradation caused by warfare

or other hostile activities.

II. FUNCTIONS

6. In the decision-making process it shall be recognized that man's needs can be met only by ensuring the proper functioning of natural systems and by respecting the principles set forth in the present Charter.

7. In the planning and implementation of social and economic development activities, due account shall be taken of the fact that the conservation of nature is an integral part of those activities.

8. In formulating long-term plans for economic development, population growth and the improvement of standards of living, due account shall be taken of the long-term capacity of natural systems to ensure the subsistence and settlement of the populations concerned, recognizing that this capacity may be enhanced through science and technology.

9. The allocation of areas of the earth to various uses shall be planned, and due account shall be taken of the physical constraints, the biological productivity and diversity and the natural beauty of the areas concerned.

10. Natural resources shall not be wasted, but used with a restraint appropriate to the principles set forth in the present Charter, in accordance with the following rules:

(a) Living resources shall not be utilized in excess of their natural capacity for regeneration;

(b) The productivity of soils shall be maintained or enhanced through measures which safeguard their long-term fertility and the process of organic decomposition, and prevent erosion and all other forms of degradation;

(c) Resources, including water, which are not consumed as they are used shall be reused or recycled;

(d) Non-renewable resources which are consumed as they are used shall be exploited with restraint, taking into account their abundance, the rational possibilities of converting them for consumption, and the compatibility of their exploitation with the functioning of natural systems.

11. Activities which might have an impact on nature shall be controlled, and the best available technologies that minimize significant risks to nature or other adverse effects shall be used; in particular:

(a) Activities which are likely to cause irreversible damage to nature shall be avoided;

(b) Activities which are likely to pose a significant risk to nature shall be preceded by an exhaustive examination; their proponents shall demonstrate that expected benefits outweigh potential damage to nature, and where potential adverse effects are not fully understood, the activities should not proceed;

(c) Activities which may disturb nature shall be preceded by assessment of their consequences, and environmental impact studies of development projects shall be conducted sufficiently in advance, and if they are to be undertaken, such activities shall be planned and carried out so as to minimize potential adverse effects;

(d) Agriculture, grazing, forestry and fisheries practices shall be adapted to the natural characteristics and constraints of given areas;

(e) Areas degraded by human activities shall be rehabilitated for purposes in accord with their natural potential and compatible with the well-being of affected populations.

12. Discharge of pollutants into natural systems shall be avoided and:

(a) Where this is not feasible, such pollutants shall be treated at the source, using the best practicable means available;
(b) Special precautions shall be taken to prevent discharge of radioactive or toxic wastes.

13. Measures intended to prevent, control or limit natural disasters, infestations and diseases shall be specifically directed to the causes of these scourges and shall avoid adverse side-effects on nature.

III. IMPLEMENTATION

14. The principles set forth in the present Charter shall be reflected in the law and practice of each State, as well as at the international level.

15. Knowledge of nature shall be broadly disseminated by all possible means, particularly by ecological education as an integral part of general education.

16. All planning shall include, among its essential elements, the formulation of strategies for the conservation of nature, the establishment of inventories of ecosystems and assessments of the effects on nature of proposed policies and activities; all of these elements shall be disclosed to the public by appropriate means in time to permit effective consultation and participation.

17. Funds, programmes and administrative structures necessary to achieve the objective of the conservation of nature shall be provided.

18. Constant efforts shall be made to increase knowledge of nature by scientific research and to disseminate such knowledge unimpeded by restrictions of any kind.

19. The status of natural processes, ecosystems and species shall

be closely monitored to enable early detection of degradation or threat, ensure timely intervention and facilitate the evaluation of conservation policies and methods.

20. Military activities damaging to nature shall be avoided.

21. States and, to the extent they are able, other public authorities, international organizations, individuals, groups and corporations shall:

(a) Co-operate in the task of conserving nature through common activities and other relevant actions, including information exchange and consultations;
(b) Establish standards for products and manufacturing processes that may have adverse effects on nature, as well as agreed methodologies for assessing these effects;
(c) Implement the applicable international legal provisions for the conservation of nature and the protection of the environment;
(d) Ensure that activities within their jurisdictions or control do not cause damage to the natural systems located within other States or in the areas beyond the limits of national jurisdiction;
(e) Safeguard and conserve nature in areas beyond national jurisdiction.

22. Taking fully into account the sovereignty of States over their natural resources, each State shall give effect to the provisions of the present Charter through its competent organs and in co-operation with other States.

23. All persons, in accordance with their national legislation, shall have the opportunity to participate, individually or with others, in the formulation of decisions of direct concern to their environment, and shall have access to means of redress when their environment has suffered damage or degradation.

24. Each person has a duty to act in accordance with the provisions of the present Charter; acting individually, in association with others or through participation in the political process, each person shall strive to ensure that the objectives and requirements of the present Charter are met.

3. Rio Declaration on Environment and Development (1992)

Date: 12 June 1992
Link: www.unep.org

Preamble

The United Nations Conference on Environment and Development,

Having met at Rio de Janeiro from 3 to 14 June 1992,

Reaffirming the Declaration of the United Nations Conference on the Human Environment, adopted at Stockholm on 16 June 1972, and seeking to build upon it, with the goal of establishing a new and equitable global partnership through the creation of new levels of cooperation among States, key sectors of societies and people,

Working towards international agreements which respect the interests of all and protect the integrity of the global environmental and developmental system,

Recognizing the integral and interdependent nature of the Earth, our home,

Proclaims that:

Principle 1

Human beings are at the centre of concerns for sustainable development. They are entitled to a healthy and productive life in harmony with nature.

Principle 2

States have, in accordance with the Charter of the United Nations and the principles of international law, the sovereign right to exploit their own resources pursuant to their own environmental and developmental policies, and the responsibility to ensure that activities within their jurisdiction or control do not cause damage to the environment of other States or of areas beyond the limits of national jurisdiction.

Principle 3

The right to development must be fulfilled so as to equitably meet developmental and environmental needs of present and future generations.

Principle 4

In order to achieve sustainable development, environmental protection shall constitute an integral part of the development process and cannot be considered in isolation from it.

Principle 5

All States and all people shall cooperate in the essential task of eradicating poverty as an indispensable requirement for sustainable development, in order to decrease the disparities in standards of living and better meet the needs of the majority of the people of the world.

Principle 6

The special situation and needs of developing countries, particularly the least developed and those most environmentally vulnerable, shall be given special priority. International actions in the field of environment and development should also address the interests and needs of all countries.

Principle 7

States shall cooperate in a spirit of global partnership to conserve, protect and restore the health and integrity of the Earth's ecosystem. In view of the different contributions to global environmental degradation, States have common but differentiated responsibilities. The developed countries acknowledge the responsibility that they bear in the international pursuit to sustainable development in view of the pressures their societies place on the global environment and of the technologies and financial resources they command.

Principle 8

To achieve sustainable development and a higher quality of life for all people, States should reduce and eliminate unsustainable patterns of production and consumption and promote appropriate demographic policies.

Principle 9

States should cooperate to strengthen endogenous capacity-building for sustainable development by improving scientific understanding through exchanges of scientific and technological knowledge, and by enhancing the development, adaptation, diffusion and transfer of technologies, including new and innovative technologies.

Principle 10

Environmental issues are best handled with participation of all concerned citizens, at the relevant level. At the national level, each individual shall have appropriate access to information concerning the environment that is held by public authorities, including information on hazardous materials and activities in their communities, and the opportunity to participate in decision-making processes. States shall facilitate and encourage public awareness and participation by making information widely available. Effective access to judicial and administrative proceedings, including redress and remedy, shall be provided.

Principle 11

States shall enact effective environmental legislation. Environmental standards, management objectives and priorities should reflect the environmental and development context to which they apply. Standards applied by some countries may be inappropriate and of unwarranted economic and social cost to other countries, in particular developing countries.

Principle 12

States should cooperate to promote a supportive and open international economic system that would lead to economic growth and sustainable development in all countries, to better address the problems of environmental degradation. Trade policy measures for environmental purposes should not constitute a means of arbitrary or unjustifiable discrimination or a disguised restriction on international trade. Unilateral actions to deal with environmental challenges outside the jurisdiction of the importing country should be avoided. Environmental measures addressing transboundary or global environmental problems should, as far as possible, be based on an international consensus.

Principle 13

States shall develop national law regarding liability and compensation for the victims of pollution and other environmental damage. States shall also cooperate in an expeditious and more determined manner to develop further international law regarding liability and compensation for adverse effects of environmental damage caused by activities within their jurisdiction or control to areas beyond their jurisdiction.

Principle 14

States should effectively cooperate to discourage or prevent the relocation and transfer to other States of any activities and substances that cause severe environmental degradation or are found to be harmful to human health.

Principle 15

In order to protect the environment, the precautionary approach shall be widely applied by States according to their capabilities. Where there are threats of serious or irreversible damage, lack of full scientific certainty shall not be used as a reason for postponing cost-effective measures to prevent environmental degradation.

Principle 16

National authorities should endeavour to promote the internalization of environmental costs and the use of economic instruments, taking into account the approach that the polluter should, in principle, bear the cost of pollution, with due regard to the public interest and without distorting international trade and investment.

Principle 17

Environmental impact assessment, as a national instrument, shall be undertaken for proposed activities that are likely to have a significant adverse impact on the environment and are subject to a decision of a competent national authority.

Principle 18

States shall immediately notify other States of any natural disasters or other emergencies that are likely to produce sudden harmful effects on the environment of those States. Every effort shall be made by the international community to help States so afflicted.

Principle 19

States shall provide prior and timely notification and relevant information to potentially affected States on activities that may have a significant adverse transboundary environmental effect and shall consult with those States at an early stage and in good faith.

Principle 20

Women have a vital role in environmental management and development. Their full participation is therefore essential to achieve sustainable development.

Principle 21

The creativity, ideals and courage of the youth of the world should be mobilized to forge a global partnership in order to achieve sustainable development and ensure a better future for all.

Principle 22

Indigenous people and their communities and other local communities have a vital role in environmental management and development because of their knowledge and traditional practices. States should recognize and duly support their identity, culture and interests and enable their effective participation in the achievement of sustainable development.

Principle 23

The environment and natural resources of people under oppression, domination and occupation shall be protected.

Principle 24

Warfare is inherently destructive of sustainable development. States shall therefore respect international law providing protection for the environment in times of armed conflict and cooperate in its further development, as necessary.

Principle 25

Peace, development and environmental protection are interdependent and indivisible.

Principle 26

States shall resolve all their environmental disputes peacefully and by appropriate means in accordance with the Charter of the United Nations.

Principle 27

States and people shall cooperate in good faith and in a spirit of partnership in the fulfillment of the principles embodied in this Declaration and in the further development of international law in the field of sustainable development.

4. Draft Articles on Prevention of Transboundary Harm from Hazardous Activities (2001)

Date: 25 January 2001
(International Law Commission 53th session)
Link: www.un.org/law/ilc

The States Parties,

Having in mind Article 13, paragraph 1 (a), of the Charter of the United Nations, which provides that the General Assembly shall initiate studies and make recommendations for the purpose of encouraging the progressive development of international law and its codification,

Bearing in mind the principle of permanent sovereignty of States over the natural resources within their territory or otherwise under their jurisdiction or control,

Bearing also in mind that the freedom of States to carry on or permit activities in their territory or otherwise under their jurisdiction or control is not unlimited,

Recalling the Rio Declaration on Environment and Development of 13 June 1992,

Recognizing the importance of promoting international co-operation,

Have agreed as follows:

Article 1
Scope

The present articles apply to activities not prohibited by international law which involve a risk of causing significant transboundary harm through their physical consequences.

Article 2
Use of terms

For the purposes of the present articles:

(a) "Risk of causing significant transboundary harm" includes risks taking the form of a high probability of causing significant transboundary harm and a low probability of causing disastrous transboundary harm;
(b) "Harm" means harm caused to persons, property or the environment;
(c) "Transboundary harm" means harm caused in the territory of or in other places under the jurisdiction or control of a State other than the State of origin, whether or not the States concerned share a common border;
(d) "State of origin" means the State in the territory or otherwise under the jurisdiction or control of which the activities referred to in article 1 are planned or are carried out;
(e) "State likely to be affected" means the State or States in the territory of which there is the risk of significant transboundary harm or which have jurisdiction or control over any other place where there is such a risk;
(f) "States concerned" means the State of origin and the State likely to be affected.

Article 3
Prevention

The State of origin shall take all appropriate measures to prevent significant transboundary harm or at any event to minimize the risk thereof.

Article 4
Cooperation

States concerned shall cooperate in good faith and, as necessary, seek the assistance of one or more competent international organizations in preventing significant transboundary harm or at any event in minimizing the risk thereof.

Article 5
Implementation

States concerned shall take the necessary legislative, administrative or other action, including the establishment of suitable monitoring mechanisms to implement the provisions of the present articles.

Article 6
Authorization

1. The State of origin shall require its prior authorization for:

(a) Any activity within the scope of the present articles carried out in its territory or otherwise under its jurisdiction or control;
(b) Any major change in an activity referred to in subparagraph (a);
(c) Any plan to change an activity which may transform it into one falling within the scope of the present articles.

2. The requirement of authorization established by a State shall be made applicable in respect of all pre-existing activities within the scope of the present articles. Authorizations already issued by the State for pre-existing activities shall be reviewed in order to comply with the present articles.

3. In case of a failure to conform to the terms of the authorization, the State of origin shall take such actions as appropriate, including where necessary terminating the authorization.

Article 7
Assessment of risk

Any decision in respect of the authorization of an activity within the scope of the present articles shall, in particular, be based on an assessment of the possible transboundary harm caused by that activity, including any environmental impact assessment.

Article 8
Notification and information

1. If the assessment referred to in article 7 indicates a risk of causing significant transboundary harm, the State of origin shall provide the State likely to be affected with timely notification of the risk and the assessment and shall transmit to it the available technical and all other relevant information on which the assessment is based.

2. The State of origin shall not take any decision on authorization of the activity pending the receipt, within a period not exceeding six months, of the response from the State likely to be affected.

Article 9
Consultations on preventive measures

1. The States concerned shall enter into consultations, at the request of any of them, with a view to achieving acceptable solutions regarding measures to be adopted in order to prevent significant transboundary harm or at any event to minimize the risk thereof. The States concerned shall agree, at the commencement of such consultations, on a reasonable time frame for the consultations.

2. The States concerned shall seek solutions based on an equitable

balance of interests in the light of article 10.

3. If the consultations referred to in paragraph 1 fail to produce an agreed solution, the State of origin shall nevertheless take into account the interests of the State likely to be affected in case it decides to authorize the activity to be pursued, without prejudice to the rights of any State likely to be affected.

Article 10
Factors involved in an equitable balance of interests

In order to achieve an equitable balance of interests as referred to in paragraph 2 of article 9, the States concerned shall take into account all relevant factors and circumstances, including:

(a) The degree of risk of significant transboundary harm and of the availability of means of preventing such harm, or minimizing the risk thereof or repairing the harm;

(b) The importance of the activity, taking into account its overall advantages of a social, economic and technical character for the State of origin in relation to the potential harm for the State likely to be affected;

(c) The risk of significant harm to the environment and the availability of means of preventing such harm, or minimizing the risk thereof or restoring the environment;

(d) The degree to which the State of origin and, as appropriate, the State likely to be affected are prepared to contribute to the costs of prevention;

(e) The economic viability of the activity in relation to the costs of prevention and to the possibility of carrying out the activity elsewhere or by other means or replacing it with an alternative activity;

(f) The standards of prevention which the State likely to be affected applies to the same or comparable activities and the standards applied in comparable regional or international practice.

Article 11
Procedures in the absence of notification

1. If a State has reasonable grounds to believe that an activity planned or carried out in the State of origin may involve a risk of causing significant transboundary harm to it, it may request the State of origin to apply the provision of article 8. The request shall be accompanied by a documented explanation setting forth its grounds.

2. In the event that the State of origin nevertheless finds that it is not under an obligation to provide a notification under article 8, it shall so inform the requesting State within a reasonable time, providing a documented explanation setting forth the reasons for such finding. If this finding does not satisfy that State, at its request, the two States shall promptly enter into consultations in the manner indicated in article 9.

3. During the course of the consultations, the State of origin shall, if so requested by the other State, arrange to introduce appropriate and feasible measures to minimize the risk and, where appropriate, to suspend the activity in question for a reasonable period.

Article 12
Exchange of information

While the activity is being carried out, the States concerned shall exchange in a timely manner all available information concerning that activity relevant to preventing significant transboundary harm or at any event minimizing the risk thereof. Such an exchange of information shall continue until such time as the States concerned consider it appropriate even after the activity is terminated.

Article 13
Information to the public

States concerned shall, by such means as are appropriate, provide the public likely to be affected by an activity within the scope of the present articles with relevant information relating to that activity, the risk involved and the harm which might result and ascertain their views.

Article 14
National security and industrial secrets

Data and information vital to the national security of the State of origin or to the protection of industrial secrets or concerning intellectual property may be withheld, but the State of origin shall cooperate in good faith with the State likely to be affected in providing as much information as possible under the circumstances.

Article 15
Non-discrimination

Unless the States concerned have agreed otherwise for the protection of the interests of persons, natural or juridical, who may be or are exposed to the risk of significant transboundary harm as a result of an activity within the scope of the present articles, a State shall not discriminate on the basis of nationality or residence or place where the injury might occur, in granting to such persons, in accordance with its legal system, access to judicial or other procedures to seek protection or other appropriate redress.

Article 16
Emergency preparedness

The State of origin shall develop contingency plans for responding to emergencies, in cooperation, where appropriate, with the State likely to be affected and competent international organizations.

Article 17
Notification of an emergency

The State of origin shall, without delay and by the most expeditious means, at its disposal, notify the State likely to be affected of an emergency concerning an activity within the scope of the present articles and provide it with all relevant and available information.

Article 18
Relationship to other rules of international law

The present articles are without prejudice to any obligation incurred by States under relevant treaties or rules of customary international law.

Article 19
Settlement of disputes

1. Any dispute concerning the interpretation or application of the present articles shall be settled expeditiously through peaceful means of settlement chosen by mutual agreement of the parties to the dispute, including negotiations, mediation, conciliation, arbitration or judicial settlement.

2. Failing an agreement on the means for the peaceful settlement of the dispute within a period of six months, the parties to the dispute shall, at the request of any of them, have recourse to the establishment of an impartial fact-finding commission.

3. The Fact-finding Commission shall be composed of one member nominated by each party to the dispute and, in addition,

a member not having the nationality of any of the parties to the dispute chosen by the nominated members who shall serve as Chairperson.

4. If more than one State is involved on one side of the dispute and those States do not agree on a common member of the Commission and each of them nominates a member, the other party to the dispute has the right to nominate an equal number of members of the Commission.

5. If the members nominated by the parties to the dispute are unable to agree on a Chairperson within three months of the request for the establishment of the Commission, any party to the dispute may request the Secretary-General of the United Nations to appoint the Chairperson who shall not have the nationality of any of the parties to the dispute. If one of the parties to the dispute fails to nominate a member within three months of the initial request pursuant to paragraph 2, any other party to the dispute may request the Secretary-General of the United Nations to appoint a person who shall not have the nationality of any of the parties to the dispute. The person so appointed shall constitute a single-member Commission.

6. The Commission shall adopt its report by a majority vote, unless it is a single-member Commission, and shall submit that report to the parties to the dispute setting forth its findings and recommendations, which the parties to the dispute shall consider in good faith.

Part II. Climate Change

5. Convention on Long-Range Transboundary Air Pollution (1979)

Date: 13 November 1979
In force: 16 March 1983
Contracting Parties: Signatories 32, Parties 51
Korea: Not yet
Link: www.unece.org

The Parties to the present Convention,

Determined to promote relations and cooperation in the field of environmental protection,

Aware of the significance of the activities of the United Nations Economic Commission for Europe in strengthening such relations and cooperation, particularly in the field of air pollution including long-range transport of air pollutants,

Recognizing the contribution of the Economic Commission for Europe to the multilateral implementation of the pertinent provisions of the Final Act of the Conference on Security and Cooperation in Europe,

Cognizant of the references in the chapter on environment of the Final Act of the Conference on Security and Cooperation in Europe calling for cooperation to control air pollution and its effects, including long-range transport of air pollutants, and to the development through international cooperation of an extensive programme for the monitoring and evaluation of long-range transport of air pollutants, starting with sulphur dioxide and with possible extension to other pollutants,

Considering the pertinent provisions of the Declaration of the

United Nations Conference on the Human Environment, and in particular principle 21, which expresses the common conviction that States have, in accordance with the Charter of the United Nations and the principles of international law, the sovereign right to exploit their own resources pursuant to their own environmental policies, and the responsibility to ensure that activities within their jurisdiction or control do not cause damage to the environment of other States or of areas beyond the limits of national jurisdiction,

Recognizing the existence of possible adverse effects, in the short and long term, of air pollution including transboundary air pollution,

Concerned that a rise in the level of emissions of air pollutants within the region as forecast may increase such adverse effects,

Recognizing the need to study the implications of the long-range transport of air pollutants and the need to seek solutions for the problems identified,

Affirming their willingness to reinforce active international cooperation to develop appropriate national policies and by means of exchange of information, consultation, research and monitoring to coordinate national action for combating air pollution including long-range transboundary air pollution,

Have agreed as follows:

Definitions

Article 1

For the purposes of the present Convention:

(a) "Air Pollution" means the introduction by man, directly or indirectly, of substances or energy into the air resulting in deleterious effects of such a nature as to endanger human health, harm living resources and ecosystems and material property and impair or interfere with amenities and other legitimate uses of the environment, and "air pollutants" shall be construed accordingly;

(b) "Long-range transboundary air pollution" means air pollution whose physical origin is situated wholly or in part within the area under the national jurisdiction of one State and which has adverse effects in the area under the jurisdiction of another State at such a distance that it is not generally possible to distinguish the contribution of individual emission sources or groups of sources.

Fundamental Principles

Article 2

The Contracting Parties, taking due account of the facts and problems involved, are determined to protect man and his environment against air pollution and shall endeavour to limit and, as far as possible, gradually reduce and prevent air pollution including long-range transboundary air pollution.

Article 3

The Contracting Parties, within the framework of the present Convention, shall by means of exchanges of information, consultation, research and monitoring, develop without undue delay policies and strategies which shall serve as a means of combating the discharge of air pollutants, taking into account efforts already made at national and international levels.

Article 4

The Contracting Parties shall exchange information on and review their policies, scientific activities and technical measures aimed at combating, as far as possible, the discharge of air pollutants which may have adverse effects, thereby contributing to the reduction of air pollution including long-range transboundary air pollution.

Article 5

Consultations shall be held, upon request, at an early stage between, on the one hand, Contracting Parties which are actually affected by or exposed to a significant risk of long-range transboundary air pollution and, on the other hand, Contracting Parties within which and subject to whose jurisdiction a significant contribution to long-range transboundary air pollution originates, or could originate, in connection with activities carried on or contemplated therein.

Air Quality Management

Article 6

Taking into account articles 2 to 5, the ongoing research, exchange of information and monitoring and the results thereof, the cost and effectiveness of local and other remedies and, in order to combat air pollution, in particular that originating from new or rebuilt installations, each Contracting Party undertakes to develop the best policies and strategies including air quality management systems and, as part of them, control measures compatible with balanced development, in particular by using the best available technology which is economically feasible and low- and non-waste technology.

Research and Development

Article 7

The Contracting Parties, as appropriate to their needs, shall initiate and co-operate in the conduct of research into and/or development of:

(a) Existing and proposed technologies for reducing emissions of sulphur compounds and other major air pollutants, including technical and economic feasibility, and environmental consequences;
(b) Instrumentation and other techniques for monitoring and measuring emission rates and ambient concentrations of air pollutants;
(c) Improved models for a better understanding of the transmission of long-range transboundary air pollutants;
(d) The effects of sulphur compounds and other major air pollutants on human health and the environment, including agriculture, forestry, materials, aquatic and other natural ecosystems and visibility, with a view to establishing a scientific basis for dose/effect relationships designed to protect the environment;
(e) The economic, social and environmental assessment of alternative measures for attaining environmental objectives including the reduction of long-range transboundary air pollution;
(f) Education and training programmes related to the environmental aspects of pollution by sulphur compounds and other major air pollutants.

Exchange of Information

Article 8

The Contracting Parties, within the framework of the Executive Body referred to in article 10 and bilaterally, shall, in their common interests, exchange available information on:

(a) Data on emissions at periods of time to be agreed upon, of agreed air pollutants, starting with sulphur dioxide, coming from grid-units of agreed size; or on the fluxes of agreed air pollutants, starting with sulphur dioxide, across national borders, at distances and at periods of time to be agreed upon;

(b) Major changes in national policies and in general industrial development, and their potential impact, which would be likely to cause significant changes in long-range transboundary air pollution;

(c) Control technologies for reducing air pollution relevant to long-range transboundary air pollution;

(d) The projected cost of the emission control of sulphur compounds and other major air pollutants on a national scale;

(e) Meteorological and physico-chemical data relating to the processes during transmission;

(f) Physico-chemical and biological data relating to the effects of long-range transboundary air pollution and the extent of the damage* which these data indicate can be attributed to long-range transboundary air pollution;

(g) National, sub-regional and regional policies and strategies for the control of sulphur compounds and other major air pollutants.

* The present Convention does not contain a rule on State liability as to damage.

Implementation and Further Development of the Cooperative Programme for the Monitoring and Evaluation of the Long-range Transmission of Air Pollutants in Europe

Article 9

The Contracting Parties stress the need for the implementation of the existing "Cooperative programme for the monitoring and evaluation of the long-range transmission of air pollutants in Europe" (hereinafter referred to as EMEP) and, with regard to the further development of this programme, agree to emphasize:

(a) The desirability of Contracting Parties joining in and fully implementing EMEP which, as a first step, is based on the monitoring of sulphur dioxide and related substances;
(b) The need to use comparable or standardized procedures for monitoring whenever possible;
(c) The desirability of basing the monitoring programme on the framework of both national and international programmes. The establishment of monitoring stations and the collection of data shall be carried out under the national jurisdiction of the country in which the monitoring stations are located;
(d) The desirability of establishing a framework for a co-operative environmental monitoring programme, based on and taking into account present and future national, sub-regional, regional and other international programmes;
(e) The need to exchange data on emissions at periods of time to be agreed upon, of agreed air pollutants, starting with sulphur dioxide, coming from grid-units of agreed size; or on the fluxes of agreed air pollutants, starting with sulphur dioxide, across national borders, at distances and at periods of time to be agreed upon. The method, including the model, used to determine the fluxes, as well as the method, including the model used to determine the transmission of

air pollutants based on the emissions per grid-unit, shall be made available and periodically reviewed, in order to improve the methods and the models;

(f) Their willingness to continue the exchange and periodic updating of national data on total emissions of agreed air pollutants, starting with sulphur dioxide;

(g) the need to provide meteorological and physico-chemical data relating to processes during transmission;

(h) the need to monitor chemical components in other media such as water, soil and vegetation, as well as a similar monitoring programme to record effects on health and environment;

(i) the desirability of extending the national EMEP networks to make them operational for control and surveillance purposes.

Executive Body

Article 10

1. The representatives of the Contracting Parties shall, within the framework of the Senior Advisers to ECE Governments on Environmental Problems, constitute the Executive Body of the present Convention, and shall meet at least annually in that capacity.

2. The Executive Body shall:

(a) Review the implementation of the present Convention;

(b) Establish, as appropriate, working groups to consider matters related to the implementation and development of the present Convention and to this end to prepare appropriate studies and other documentation and to submit recommendations to be considered by the Executive Body;

(c) Fulfil such other functions as may be appropriate under the provisions of the present Convention.

3. The Executive Body shall utilize the Steering Body for the EMEP to play an integral part in the operation of the present Convention, in particular with regard to data collection and scientific cooperation.

4. The Executive Body, in discharging its functions, shall, when it deems appropriate, also make use of information from other relevant international organizations.

Secretariat

Article 11

The Executive Secretary of the Economic Commission for Europe shall carry out, for the Executive Body, the following secretariat functions:

(a) To convene and prepare the meetings of the Executive Body;
(b) To transmit to the Contracting Parties reports and other information received in accordance with the provisions of the present Convention;
(c) To discharge the functions assigned by the Executive Body.

Amendment to the Convention

Article 12

1. Any Contracting Party may propose amendments to the present Convention.

2. The text of proposed amendments shall be submitted in writing to the Executive Secretary of the Economic Commission for Europe, who shall communicate them to all Contracting Parties. The Executive Body shall discuss proposed amendments at its next annual meeting provided that such proposals have been circulated by the Executive Secretary of the Economic Commission

for Europe to the Contracting Parties at least ninety days in advance.

3. An amendment to the present Convention shall be adopted by consensus of the representatives of the Contracting Parties, and shall enter into force for the Contracting Parties which have accepted it on the ninetieth day after the date on which two-thirds of the Contracting Parties have deposited their instruments of acceptance with the depositary. Thereafter, the amendment shall enter into force for any other Contracting Party on the ninetieth day after the date on which that Contracting Party deposits its instrument of acceptance of the amendment.

Settlement of Disputes

Article 13

If a dispute arises between two or more Contracting Parties to the present Convention as to the interpretation or application of the Convention, they shall seek a solution by negotiation or by any other method of dispute settlement acceptable to the parties to the dispute.

Signature

Article 14

1. The present Convention shall be open for signature at the United Nations Office at Geneva from 13 to 16 November 1979 on the occasion of the High-level Meeting within the framework of the Economic Commission for Europe on the Protection of the Environment, by the member States of the Economic Commission for Europe as well as States having consultative status with the Economic Commission for Europe, pursuant to paragraph 8 of

Economic and Social Council resolution 36(IV) of 28 March 1947, and by regional economic integration organizations, constituted by sovereign States members of the Economic Commission for Europe, which have competence in respect of the negotiation, conclusion and application of international agreements in matters covered by the present Convention.

2. In matters within their competence, such regional economic integration organizations shall, on their own behalf, exercise the rights and fulfil the responsibilities which the present Convention attributes to their member States. In such cases, the member States of these organizations shall not be entitled to exercise such rights individually.

Ratification, Acceptance, Approval and Accession

Article 15

1. The present Convention shall be subject to ratification, acceptance or approval.

2. The present Convention shall be open for accession as from 17 November 1979 by the States and organizations referred to in article 14, paragraph 1.

3. The instruments of ratification, acceptance, approval or accession shall be deposited with the Secretary-General of the United Nations, who will perform the functions of the depositary.

Entry into Force

Article 16

1. The present Convention shall enter into force on the ninetieth day after the date of deposit of the twenty-fourth instrument of

ratification, acceptance, approval or accession.

2. For each Contracting Party which ratifies, accepts or approves the present Convention or accedes thereto after the deposit of the twenty-fourth instrument of ratification, acceptance, approval or accession, the Convention shall enter into force on the ninetieth day after the date of deposit by such Contracting Party of its instrument of ratification, acceptance, approval or accession.

Withdrawal

Article 17

At any time after five years from the date on which the present Convention has come into force with respect to a Contracting Party, that Contracting Party may withdraw from the Convention by giving written notification to the depositary. Any such withdrawal shall take effect on the ninetieth day after the date of its receipt by the depositary.

Authentic Texts

Article 18

The original of the present Convention, of which the English, French and Russian texts are equally authentic, shall be deposited with the Secretary-General of the United Nations.

IN WITNESS WHEREOF the undersigned, being duly authorized thereto, have signed the present Convention.

DONE at Geneva, this thirteenth day of November, one thousand nine hundred and seventy-nine.

6. Vienna Convention for the Protection of Ozone Layer (1985)

Date: 22 March 1985
In force: 22 September 1988
Contracting Parties: Signatories 28, Parties 196
Korea: 27 May 1992 (No.1089)
Link: www.ozone.unep.org

Preamble

The Parties to this Convention,

Aware of the potentially harmful impact on human health and the environment through modification of the ozone layer,

Recalling the pertinent provisions of the Declaration of the United Nations Conference on the Human Environment, and in particular principle 21, which provides that "States have, in accordance with the Charter of the United Nations and the principles of international law, the sovereign right to exploit their own resources pursuant to their own environmental policies, and the responsibility to ensure that activities within their jurisdiction or control do not cause damage to the environment of other States or of areas beyond the limits of national jurisdiction,"

Taking into account the circumstances and particular requirements of developing countries,

Mindful of the work and studies proceeding within both international and national organizations and, in particular, of the World Plan of Action on the Ozone Layer of the United Nations Environment Programme,

Mindful also of the precautionary measures for the protection of the ozone layer which have already been taken at the national and international levels,

Aware that measures to protect the ozone layer from modifications due to human activities require international co-operation and action and should be based on relevant scientific and technical considerations,

Aware also of the need for further research and systematic observations to further develop scientific knowledge of the ozone layer and possible adverse effects resulting from its modification,

Determined to protect human health and the environment against adverse effects resulting from modifications of the ozone layer,

Have agreed as follows:

Article 1
Definitions

For the purposes of this Convention:

1. "The ozone layer" means the layer of atmospheric ozone above the planetary boundary layer.

2. "Adverse effects" means changes in the physical environment or biota, including changes in climate, which have significant deleterious effects on human health or on the composition, resilience and productivity of natural and managed ecosystems, or on materials useful to mankind.

3. "Alternative technologies or equipment" means technologies or equipment the use of which makes it possible to reduce or effectively eliminate emissions of substances which have or are likely to have adverse effects on the ozone layer.

4. "Alternative substances" means substances which reduce, eliminate or avoid adverse effects on the ozone layer.

5. "Parties" means, unless the text otherwise indicates, Parties to this Convention.

6. "Regional economic integration organization" means an organization constituted by sovereign States of a given region which has competence in respect of matters governed by this Convention or its protocols and has been duly authorized, in accordance with its internal procedures, to sign, ratify, accept, approve or accede to the instruments concerned.

7. "Protocols" means protocols to this Convention.

Article 2
General Obligations

1. The Parties shall take appropriate measures in accordance with the provisions of this Convention and of those protocols in force to which they are party to protect human health and the environment against adverse effects resulting or likely to result from human activities which modify or are likely to modify the ozone layer.

2. To this end the Parties shall, in accordance with the means at their disposal and their capabilities:

(a) Co-operate by means of systematic observations, research and information exchange in order to better understand and assess the effects of human activities on the ozone layer and the effects on human health and the environment from modification of the ozone layer

(b) Adopt appropriate legislative or administrative measures and co-operate in harmonizing appropriate policies to control, limit, reduce or prevent human activities under their. jurisdiction or control should it be found that these activities have or are likely to have adverse effects resulting from modification or likely modification of the ozone layer;

(c) Co-operate in the formulation of agreed measures, procedures and standards for the implementation of this Convention, with a view to the adoption of protocols and annexes;

(d) Co-operate with competent international bodies to implement effectively this Convention and protocols to which they are party.

3. The provisions of this Convention shall in no way affect the right of Parties to adopt, in accordance with international law, domestic measures additional to those referred to in paragraphs 1 and 2 above, nor shall they affect additional domestic measures already taken by a Party, provided that these measures are not incompatible with their obligations under this Convention.

4. The application of this Article shall be based on relevant scientific and technical considerations.

Article 3
Research and Systematic Observations

1. The Parties undertake, as appropriate, to initiate and cooperate in, directly or through competent international bodies, the conduct of research and scientific assessments on:

(a) The physical and chemical processes that may affect the ozone layer;

(b) The human health and other biological effects deriving from any modifications of the ozone layer, particularly those resulting from changes in ultra-violet solar radiation having biological effects(UV- B);

(c) Climatic effects deriving from any modifications of the ozone layer;

(d) Effects deriving from any modifications of the ozone layer and any consequent change in UV-B radiation on natural and synthetic materials useful to mankind;
(e) Substances, practices, processes and activities that may affect the ozone layer, and their cumulative effects;
(f) Alternative substances and technologies;
(g) Related socio-economic matters; and as further elaborated in annexes I and II.

2. The Parties undertake to promote or establish, as appropriate, directly or through competent international bodies and taking fully into account national legislation and relevant ongoing activities at both the national and international levels, joint or complementary programmes for systematic observation of the state of the ozone layer and other relevant parameters, as elaborated in Annex I.

3. The Parties undertake to co-operate, directly or through competent international bodies, in ensuring the collection, validation and transmission of research and observational data through appropriate world data centres in a regular and timely fashion.

Article 4
Co-operation in the Legal, Scientific and Technical Fields

1. The Parties shall facilitate and encourage the exchange of scientific, technical, socioeconomic, commercial and legal information relevant to this Convention as further elaborated in Annex II. Such information shall be supplied to bodies agreed upon by the Parties. Any such body receiving information regarded as confidential by the supplying Party shall ensure that such information is not disclosed and shall aggregate it to protect its confidentiality before it is made available to all Parties.

2. The Parties shall co-operate, consistent with their national laws,

regulations and practices and taking into account in particular the needs of the developing countries, in promoting, directly or through competent international bodies, the development and transfer of technology and knowledge. Such co-operation shall be carried out particularly through:

(a) Facilitation of the acquisition of alternative technologies by other Parties;
(b) Provision of information on alternative technologies and equipment, and supply of special manuals or guides to them;
(c) The supply of necessary equipment and facilities for research and systematic observations;
(d) Appropriate training of scientific and technical personnel.

Article 5
Transmission of Information

The Parties shall transmit, through the secretariat, to the Conference of the Parties established under Article 6 information on the measures adopted by them in implementation of this convention and of protocols to which they are party in such form and at such intervals as the meetings of the parties to the relevant instruments may determine.

Article 6
Conference of the Parties

1. A Conference of the Parties is hereby established. The first meeting of the Conference of the Parties shall be convened by the secretariat designated on an interim basis under Article 7 not later than one year after entry into force of this Convention. Thereafter, ordinary meetings of the Conference of the Parties shall be held at regular intervals to be determined by the

Conference at its first meeting.

2. Extraordinary meetings of the Conference of the Parties shall be held at such other times as may be deemed necessary by the Conference, or at the written request of any party, provided that, within six months of the request being communicated to them by the secretariat, it is supported by at least one third of the Parties.

3. The Conference of the Parties shall by consensus agree upon and adopt rules of procedure and financial rules for itself and for any subsidiary bodies it may establish, as well as financial provisions governing the functioning of the secretariat.

4. The Conference of the Parties shall keep under continuous review the implementation of this Convention, and, in addition, shall:

(a) Establish the form and the intervals for transmitting the information to be submitted in accordance with Article 5 and consider such information as well as reports submitted by any subsidiary body;
(b) Review the scientific information on the ozone layer, on its possible modification and on possible effects of any such modification;
(c) Promote, in accordance with Article 2, the harmonization of appropriate policies, strategies and measures for minimizing the release of substances causing or likely to cause modification of the ozone layer, and make recommendations on any other measures relating to this Convention;
(d) Adopt, in accordance with Articles 3 and 4, programmes for research, systematic observations, scientific and technological co-operation, the exchange of information and the transfer of technology and knowledge;
(e) Consider and adopt, as required, in accordance with Articles 9 and 10, amendments to this Convention and its annexes;

(f) Consider amendments to any protocol, as well as to any annexes thereto, and, if so decided, recommend their adoption to the parties to the concerned;
(g) Consider and adopt, as required, in accordance with article 10, additional annexes to this Convention;
(h) Consider and adopt, as required, protocols in accordance with article 8;
(i) Establish such subsidiary bodies as are deemed necessary for the implementation of this Convention;
(j) Seek, where appropriate, the services of competent international bodies and scientific committees, in particular the World Meteorological Organization and the World Health Organization, as well as the Co-coordinating Committee on the Ozone Layer, in scientific research, systematic observations and other activities pertinent to the objectives of this Convention, and make use as appropriate of information from these bodies and committees;
(k) Consider and undertake any additional action that may be required for the achievement of the purposes of this Convention.

5. The United Nations, its specialized agencies and the International Atomic Energy Agency, as well as any State not party to this Convention, may be represented at meetings of the Conference of the Parties by observers. Any body or agency, whether national or international, governmental or nongovernmental, qualified in fields relating to the protection of the ozone layer which has informed the secretariat of its wish to be represented at a meeting of the Conference of the Parties as an observer may be admitted unless at least one-third of the Parties present object. The admission and participation of observers shall be subject to the rules of procedure adopted by the Conference of the Parties.

Article 7
Secretariat

1. The functions of the secretariat shall be:

(a) To arrange for and service meetings provided for in articles 6, 8, 9 and 10;
(b) To prepare and transmit reports based upon information received in accordance with articles 4 and 5, as well as upon information derived from meetings of subsidiary bodies established under Article 6;
(c) To perform the functions assigned to it by any protocols;
(d) To prepare reports on its activities carried out in implementation of its functions under this Convention and present them to the Conference of the Parties;
(e) To ensure the necessary co-ordination with other relevant international bodies, and in particular to enter into such administrative and contractual arrangements as may be required for the effective discharge of its functions;
(f) To perform such other functions as may be determined by the Conference of the Parties.

2. The secretariat functions will be carried out on an interim basis by the United Nations Environment Programme until the completion of the first ordinary meeting of the Conference of the Parties held pursuant to Article 6. At its first ordinary meeting, the Conference of the Parties shall designate the secretariat from amongst those existing competent international organizations which has signified their willingness to carry out the secretariat functions under this Convention.

Article 8
Adoption of Protocols

1. The Conference of the Parties may at a meeting adopt

protocols pursuant to Article 2.

2. The text of any proposed protocol shall be communicated to the Parties by the secretariat at least six months before such a meeting.

Article 9
Amendment of the Convention or Protocols

1. Any Party may propose amendments to this Convention or to any protocol. Such amendments shall take due account, inter alia, of relevant scientific and technical considerations.

2. Amendments to this Convention shall be adopted at a meeting of the Conference of the Parties. Amendments to any protocol shall be adopted at a meeting of the Parties to the protocol in question. The text of any proposed amendment to this Convention or to any protocol, except as may otherwise be provided in such protocol, shall be communicated to the Parties by the secretariat at least six months before the meeting at which it is proposed for adoption. The secretariat shall also communicate proposed amendments to the signatories to this Convention for information.

3. The Parties shall make every effort to reach agreement on any proposed amendment to this Convention by consensus. If all efforts at consensus have been exhausted, and no agreement reached, the amendment shall as a last resort be adopted by a three-fourth majority vote of the Parties present and voting at the meeting, and shall be submitted by the Depositary to all Parties for ratification, approval or acceptance.

4. The procedure mentioned in paragraph 3 above shall apply to amendments to any protocol, except that a two-thirds majority of the parties to that protocol present and voting at the meeting shall suffice for their adoption.

5. Ratification, approval or acceptance of amendments shall be notified to the Depositary in writing. Amendments adopted in accordance with paragraphs 3 or 4 above shall enter into force between parties having accepted them on the ninetieth day after the receipt by the Depositary of notification of their ratification, approval or acceptance by at least three- fourths of the Parties to this Convention or by at least two-thirds of the parties to the protocol concerned, except as may otherwise be provided in such protocol. Thereafter the amendments shall enter into force for any other Party on the ninetieth day after that Party deposits its instrument of ratification, approval or acceptance of the amendments.

6. For the purposes of this Article, "Parties present and voting" means Parties present and casting an affirmative or negative vote.

Article 10
Adoption and Amendment of Annexes

1. The annexes to this Convention or to any protocol shall form an integral part of this Convention or of such protocol, as the case may be, and, unless expressly provided otherwise, a reference to this Convention or its protocols constitutes at the same time a reference to any annexes thereto. Such annexes shall be restricted to scientific, technical and administrative matters.

2. Except as may be otherwise provided in any protocol with respect to its annexes, the following procedure shall apply to the proposal, adoption and entry into force of additional annexes to this Convention or of annexes to a protocol;

(a) Annexes to this Convention shall be proposed and adopted according to the procedure laid down in Article 9, paragraphs 2 and 3, while annexes to any protocol shall be proposed and adopted according to the procedure laid down in Article 9, paragraphs 2 and 4;

(b) Any party that is unable to approve an additional annex to this Convention or an annex to any protocol to which it is party shall so notify the Depositary, in writing, within six months from the date of the communication of the adoption by the Depositary. The Depositary shall without delay notify all Parties of any such notification received. A Party may at any time substitute an acceptance for a previous declaration of objection and the annexes shall thereupon enter into force for that Party;

(c) On the expiry of six months from the date of the circulation of the communication by the Depositary, the annex shall become effective for all Parties to this Convention or to any protocol concerned which have not submitted a notification in accordance with the provision of subparagraph (b) above.

3. The proposal, adoption and entry into force of amendments to annexes to this Convention or to any protocol shall be subject to the same procedure as for the proposal, adoption and entry into force of annexes to the Convention or annexes to a protocol. Annexes and amendments thereto shall take due account, inter alia, of relevant scientific and technical considerations.

4. If an additional annex or an amendment to an annex involves an amendment to this Convention or to any protocol, the additional annex or amended annex shall not enter into force until such time as the amendment to this Convention or to the protocol concerned enters into force.

Article 11
Settlement of Disputes

1. In the event of a dispute between Parties concerning the interpretation or application of this Convention, the parties

concerned shall seek solution by negotiation.

2. If the parties concerned cannot reach agreement by negotiation, they may jointly seek the good offices of, or request mediation by, a third party.

3. When ratifying, accepting, approving or acceding to this Convention, or at any time thereafter, a State or regional economic integration organization may declare in writing to the Depositary that for a dispute not resolved in accordance with paragraph 1 or paragraph 2 above, it accepts one or both of the following means of dispute settlement as compulsory:

(a) Arbitration in accordance with procedures to be adopted by the Conference of the Parties at its first ordinary meeting;
(b) Submission of the dispute to the International Court of Justice.

4. If the parties have not, in accordance with paragraph 3 above, accepted the same or any procedure, the dispute shall be submitted to conciliation in accordance with paragraph 5 below unless the parties otherwise agree.

5. A conciliation commission shall be created upon the request of one of the parties to the dispute. The commission shall be composed of an equal number of members appointed by each party concerned and a chairman chosen jointly by the members appointed by each party. The commission shall render a final and recommendatory award, which the parties shall consider in good faith.

6. The provisions of this article shall apply with respect to any protocol except as otherwise provided in the protocol concerned.

Article 12
Signature

This Convention shall be open for signature at the Federal Ministry for Foreign Affairs of the Republic of Austria in Vienna from 22 March 1985 to 21 September 1985, and at United Nations Headquarters in New York from 22 September 1985 to 21 March 1986.

Article 13
Ratification, Acceptance or Approval

1. This Convention and any protocol shalt be subject to ratification, acceptance or approval by States and by regional economic integration organizations. Instruments of ratification, acceptance or approval shall be deposited with the Depositary.

2. Any organization referred to in paragraph 1 above which becomes a Party to this Convention or any protocol without any of its member States being a Party shall be bound by all the obligations under the Convention or the protocol as the case may be. In the case of such organizations, one or more of whose member States is a Party to the Convention or relevant protocol, the organization and its member Stales shall decide on their respective responsibilities for the performance of their obligation under the Convention or protocol, as the case may be. In such cases, the organization and the member States shall not be entitled to exercise rights under the Convention or relevant protocol concurrently.

3. In their instruments of ratification, acceptance or approval, the organizations referred to in paragraph 1 above shall declare the extent of their competence with respect to the matters governed by the Convention or the relevant protocol. These organizations shall also inform the Depositary of any substantial modification in

the extent of their competence.

Article 14
Accession

1. This Convention and any protocol shall be open for accession by States and by regional economic integration organizations from the date on which the Convention or the protocol concerned is closed for signature. The instruments of accession shall be deposited with the Depositary.

2. In their instruments of accession, the organizations referred to in paragraph 1 above shall declare the extent of their competence with respect to the matters governed by the Convention or the relevant protocol. These organizations shall also inform the Depositary of any substantial modification in the extent of their competence.

3. The provisions of article 13, paragraph 2, shall apply to regional economic integration organizations which accede to this Convention or any protocol.

Article 15
Right to Vote

1. Each Party to this Convention or to any protocol shall have one vote.

2. Except as provided for in paragraph 1 above, regional economic integration organizations, in matters within their competence, shall exercise their right to vote with a number of votes equal to the number of their member States which are Parties to the Convention or the relevant protocol. Such organizations shall be exercise their right to vote if their member States exercise theirs, and vice versa.

Article 16
Relationship Between the Convention and its Protocols

1. A State or a regional economic integration organization may not become a party to a protocol unless it is, or becomes at the same time, a Party to the Convention.

2. Decisions concerning any protocol shall be taken only by the parties to the protocol concerned.

Article 17
Entry into Force

1. This Convention shall enter into force on the ninetieth day after the date of deposit of the twentieth instrument of ratification, acceptance, approval or accession.

2. Any protocol, except as otherwise provided in such protocol, shall enter into force on the ninetieth day after the date of deposit of the eleventh instrument of ratification, acceptance or approval of such protocol or accession thereto.

3. For each Party which ratifies, accepts or approves this Convention or accedes thereto after the deposit of the twentieth instrument of ratification, acceptance, approval or accession, it shall enter into force on the ninetieth day after the date of deposit by such Party of its instrument of ratification, acceptance, approval or accession.

4. Any protocol, except as otherwise provided in such protocol, shall enter into force for a party that ratifies, accepts or approves that protocol or accedes thereto after its entry into force pursuant to paragraph 2 above, on the ninetieth day after the date on which that party deposits its instrument of ratification, acceptance, approval or accession, or on the date on which the Convention enters into force for that Party, whichever shall be the later.

5. For the purposes of paragraphs 1 and 2 above, any instrument deposited by a regional economic integration organization shall not be counted as additional to those deposited by member Sacs of such organization.

Article 18
Reservations

No reservations may be made to this Convention.

Article 19
Withdrawal

1. At any time after four years from the date on which this Convention has entered into force for a Party that Party may withdraw from the Convention by giving written notification to the Depositary.

2. Except as may be provided in any protocol, at any time after four years from the date on which such protocol has entered into force for a party that party may withdraw from the protocol by giving written notification to the Depositary.

3. Any such withdrawal shall take effect upon expiry of one year after the date of its receipt by the Depositary, or on such later date as may be specified in the notification of the withdrawal.

4. Any Party which withdraws from this Convention shall be considered as also having withdrawn from any protocol to which it is party.

Article 20
Depositary

1. The Secretary-General of the United Nations shall assume the

functions of depositary of this Convention and any protocols.

2. The Depositary shall inform the Parties, in particular, of:

(a) The signature of this Convention and of any protocol, and the deposit of instruments of ratification, acceptance, approval or accession in accordance with articles 13 and 14;
(b) The date on which the Convention and any protocol will come into force in accordance with article 17;
(c) Notifications of withdrawal made in accordance with article 19;
(d) Amendments adopted with respect to the Convention and any protocol, their acceptance by the parties and their date of entry into force in accordance with Article 9;
(e) All communications relating to the adoption and approval of annexes and to the amendment of annexes in accordance with article 10;
(f) Notifications by regional economic integration organizations of the extent of their competence with respect to matters governed by this Convention and any protocols, and of any modifications thereof.
(g) Declarations made in accordance with article 11, paragraph 3.

Article 21
Authentic Texts

The original of this Convention, of which the Arabic, Chinese, English, French, Russian and Spanish texts are equally authentic, shall be deposited with the Secretary-General of the United Nations.

IN WITNESS WHEREOF the undersigned, being duly authorized to that effect, have signed this Convention.

DONE at Vienna on the 22nd day of March 1985

Annex I
Research and Systematic Observations

1. The Parties to the Convention recognize that the major scientific issues are:

(a) Modification of the ozone layer which would result in a change in the amount of solar ultra- violet radiation having biological effects(UV-B) that reaches the Earth's surface and the potential consequences for human health, for organisms, ecosystems and materials useful to mankind;

(b) Modification of the vertical distribution of ozone, which could change the temperature structure of the atmosphere and the potential consequences for weather and climate.

2. The Parties to the Convention, in accordance with Article 3, shall co-operate in conducting research and systematic observations and in formulating recommendations for future research and observation in such areas as:

(a) Research into the physics and chemistry of the atmosphere

(i) Comprehensive theoretical models: further development of models which consider the interaction between radiative, dynamic and chemical processes; studies of the simultaneous effects of various man-made and naturally occurring species upon atmospheric ozone; interpretation of satellite and non-satellite measurement data sets; evaluation of trends in atmospheric and geophysical parameters, and the development of methods for attributing changes in these parameters to specific causes;

(ii) Laboratory studies of: rate coefficients, absorption cross-sections and mechanisms of tropospheric and stratospheric chemical and photochemical processes; spectroscopic data to support field measurements in all relevant spectral regions;

(iii) Field measurements: the concentration and fluxes of key source gases of both natural and anthropogenic origin; atmospheric dynamics studies; simultaneous measurements of photochemically- related species down to the planetary boundary layer, using in situ and remote sensing instruments; intercomparison of different sensors, including co-ordinated correlative measurements for satellite instrumentation; three-dimensional fields of key atmospheric trace constituents, solar spectral flux and meteoro- logical parameters;

(iv) Instrument development, including satellite and non-satellite sensors for atmospheric trace constituents, solar flux and meteorological parameters

(b) Research into health, biological and photodegradation effects

(i) The relationship between human exposure to visible and ultra-violet solar radiation and (a) the development of both non-melanoma and melanoma skin cancer and (b) the effects on the immunological system;

(ii) Effects of UV-B radiation, including the wavelength dependence, upon (a) agricultural crops, forests and other terrestrial ecosystems and (b) the aquatic food web and fisheries, as well as possible inhibition of oxygen production by marine phytoplankton;

(iii) The mechanisms by which UV-B radiation acts on biological materials, species and ecosystems, including: the relationship between dose, dose rate, and response; photo-repair, adaptation, and protection;

(iv) Studies of biological action spectra and the spectral response using polychromatic radiation in order to include possible interactions of the various wavelength regions;

(v) The influence of UV-B radiation on: the sensitivities and activities of biological species important to the biospheric balance; primary processes such as photosynthesis and biosynthesis;

(vi) The influence of UV-B radiation on the photo-degradation of pollutants, agricultural chemicals and other materials;

(c) Research on effects on climate

(i) Theoretical and observational studies of the radiative effects of ozone and other trace species and the impact on climate parameters, such as land and ocean surface temperatures, precipitation patterns, the exchange between the troposphere and stratosphere;

(ii) The investigation of the effects of such climate impacts on various aspects of human activity;

(d) Systematic observations on:

(i) The status of the ozone layer(i.e. the spatial and temporal variability of the total column content and vertical distribution) by making the Global Ozone Observing System, based on the integration of satellite and ground-based systems, fully operational;

(ii) The tropospheric and stratospheric concentrations of source gases for the HOx, NOx, ClOx and carbon families; (iii) The temperature from the ground to the mesosphere, utilizing both ground-based and satellite systems;

(iv) Wavelength-resolved solar flux reaching, and thermal radiation leaving, the Earth's atmosphere, utilizing satellite measurements;

(v) Wavelength-resolved solar flux reaching the Earth's surface in the ultra-violet range having biological effects (UV-B);

(vi) Aerosol properties and distribution from the ground to the mesosphere, utilizing ground-based, airborne and satellite systems;

(vii) Climatically important variables by the maintenance of programmes of high-quality meteorological surface measurements;

(viii) Trace species, temperatures, solar flux and aerosols utilizing improved methods for analyzing global data.

3. The Parties to the Convention shall co-operate, taking into account the particular needs of the developing countries, in promoting the appropriate scientific and technical training required to participate in the research and systematic observations outlined in this annex. Particular emphasis should be given to the inter-calibration of observational instrumentation and methods with a view to generating comparable or standardized scientific data sets.

4. The following chemical substances of natural and anthropogenic origin, not listed in order of priority, are thought to have the potential to modify the chemical and physical properties of the ozone layer.

(a) Carbon substances

(i) Carbon monoxide(CO)

Carbon monoxide has significant natural and anthropogenic sources, and is thought to play a major direct role in tropospheric photochemistry, and an indirect role in stratospheric photochemistry.

(ii) Carbon dioxide(CO2)

Carbon dioxide has significant natural and anthropogenic sources, and affects stratospheric ozone by influencing the thermal structure of the atmosphere.

(iii) Methane(CH4)

Methane has both natural and anthropogenic sources, and affects both tropospheric and stratospheric ozone.

(iv) Non-methane hydrocarbon species

Non-methane hydrocarbon species, which consist of a large number of chemical substances, have both natural and anthropogenic sources, and play a direct role in tropospheric photochemistry and an indirect role in stratospheric photochemistry.

(b) Nitrogen substances

(i) Nitrous oxide(N_2O)

The dominant sources of N_2O are natural, but anthropogenic contributions are becoming increasingly important. Nitrous oxide is the primary source of stratospheric NOx, which play a vital role in controlling the abundance of stratospheric ozone.

(ii) Nitrogen oxides(NOx)

round-level sources of NOx play a major direct role only in tropospheric photochemical processes and an indirect role in stratosphere photochemistry, whereas injection of NOx close to the tropopause may lead directly to a change in upper tropospheric and stratospheric ozone.

(c) Chlorine substances

(i) Fully halogenated alkanes, e.g. CCl_4, $CFCl_3$(CFC-11), CF_2Cl_2(CFC-12), $C_2F_3Cl_3$(CFC- 113), $C_2F_4Cl_2$(CFC-114)

Fully halogenated alkanes are anthropogenic and act as a source of ClOx, which plays a vital role in ozone photochemistry, especially in the 30-50 km altitude region.

(ii) Partially halogenated alkanes, e.g. CH_3Cl, CHF_2Cl (CFC-22), CH_3CCl_3, $CHFCl_2$(CFC- 21)

The sources of CH_3Cl are natural, whereas the other partially halogenated alkanes mentioned above are anthropogenic in origin.

These gases also act as a source of stratospheric ClOx.

(d) Bromine substances

Fully halogenated alkanes, e.g. CF_3Br

These gases are anthropogenic and act as a source of BrOx, which behaves in a manner similar to ClOx.

(e) Hydrogen substances

(i) Hydrogen(H_2)

Hydrogen, the source of which is natural and anthropogenic, plays a minor role in stratospheric photochemistry.

(ii) Water(H_2O)

Water, the source of which is natural, plays a vital role in both tropospheric and stratospheric photochemistry. Local sources of water vapour in the stratosphere include the oxidation of methane and, to a lesser extent, of hydrogen.

Annex II
Information Exchange

1. The Parties to the Convention recognize that the collection and sharing of information is an important means of implementing the objectives of this Convention and of assuring that any actions that may be taken are appropriate and equitable. Therefore, Parties shall exchange scientific, technical, socio-economic, business, commercial and legal information.

2. The Parties to the Convention, in deciding what information is to be collected and exchanged, should take into account the usefulness of the information and the costs of obtaining it. The Parties further recognize that co-operation under this annex has to be consistent with national laws, regulations and practices regarding patents, trade secrets, and protection of confidential and proprietary information.

3. Scientific information

This includes information on:

(a) Planned and ongoing research, both governmental and private, to facilitate the co-ordination of research programmes so as to make the most effective use of available national and international resources;

(b) The emission data needed for research;

(c) Scientific results published in peer-reviewed literature on the understanding of the physics and chemistry of the Earth's atmosphere and of its susceptibility to change, in particular on the state of the ozone layer and effects on human health, environment and climate which would result from changes on all time-scales in either the total column content or the vertical distribution of ozone;

(d) The assessment of research results and the recommendations for future research.

4. Technical information

This includes information on:

(a) The availability and cost of chemical substitutes and of alternative technologies to reduce the emissions of ozone-modifying substances and related planned and ongoing research;

(b) The limitations and any risks involved in using chemical or other substitutes and alternative technologies.

5. Socio-economic and commercial information on the substances referred to in Annex I

This includes information on:

(a) Production and production capacity;

(b) Use and use patterns;

(c) Imports/exports;

(d) The costs, risks and benefits of human activities which may indirectly modify the ozone layer and of the impacts of regulatory actions taken or being considered to control these activities.

6. Legal information

This includes information on:

(a) National laws, administrative measures and legal research relevant to the protection of the ozone layer;
(b) International agreements, including bilateral agreements, relevant to the protection of the ozone layer;
(c) Methods and terms of licensing and availability of patents relevant to the protection of the ozone layer.

7. Montreal Protocol on Substances that Deplete the Ozone Layer (1987)

Date: 16 September 1987
In force: 1 January 1989
Contracting Parties: Signatories 46, Parties 196
Korea: 27 May 1992 (No.1090)
Link: www.ozone.unep.org

Introduction to the adjustments

The Second (1990), Fourth (1992), Seventh (1995), Ninth (1997), Eleventh (1999) and Nineteenth (2007) Meetings of the Parties to the Montreal Protocol on Substances that Deplete the Ozone Layer decided, on the basis of assessments made pursuant to Article 6 of the Protocol, to adopt adjustments and reductions of production and consumption of the controlled substances in Annexes A, B, C and E to the Protocol as follows (the text here shows the cumulative effect of all the adjustments):

Preamble

The Parties to this Protocol,

Being Parties to the Vienna Convention for the Protection of the Ozone Layer,

Mindful of their obligation under that Convention to take appropriate measures to protect human health and the environment against adverse effects resulting or likely to result from human activities which modify or are likely to modify the ozone layer,

Recognizing that world-wide emissions of certain substances can significantly deplete and otherwise modify the ozone layer in a manner that is likely to result in adverse effects on human health and the environment,

Conscious of the potential climatic effects of emissions of these substances,

Aware that measures taken to protect the ozone layer from depletion should be based on relevant scientific knowledge, taking into account technical and economic considerations,

Determined to protect the ozone layer by taking precautionary measures to control equitably total global emissions of substances that deplete it, with the ultimate objective of their elimination on the basis of developments in scientific knowledge, taking into account technical and economic considerations,

Acknowledging that special provision is required to meet the needs of developing countries for these substances,

Noting the precautionary measures for controlling emissions of certain chlorofluorocarbons that have already been taken at national and regional levels,

Considering the importance of promoting international co-operation in the research and development of science and technology relating to the control and reduction of emissions of substances that deplete the ozone layer, bearing in mind in particular the needs of developing countries,

Have Agreed As Follows:

Article 1
Definitions

For the purposes of this Protocol:

1. "Convention" means the Vienna Convention for the Protection of the Ozone Layer, adopted on 22 March 1985.

2. "Parties" means, unless the text otherwise indicates, Parties to this Protocol.

3. "Secretariat" means the Secretariat of the Convention.

4. "Controlled substance" means a substance listed in Annex A to this Protocol, whether existing alone or in a mixture. It excludes, however, any such substance or mixture which is in a manufactured product other than a container used for the transportation or storage of the substance listed.

5. "Production" means the amount of controlled substances produced, minus the amount destroyed by technologies to be approved by the Parties.

6. "Consumption" means production plus imports minus exports of controlled substances.

7. "Calculated levels" of production, imports, exports and consumption means levels determined in accordance with Article 3.

8. "Industrial rationalization" means the transfer of all or a portion of the calculated level of production of one Party to another, for the purpose of achieving economic efficiencies or responding to anticipated shortfalls in supply as a result of plant closures.

Article 2
Control Measures

1. *Incorporated in Article 2A.*

2. *Replaced by Article 2B.*

3. *Replaced by Article 2A.*

4. *Replaced by Article 2A.*

5. Any Party may, for one or more control periods, transfer to another Party any portion of its calculated level of production set out in Articles 2A to 2F, and Article 2H, provided that the total combined calculated levels of production of the Parties concerned for any group of controlled substances do not exceed the production limits set out in those Articles for that group. Such transfer of production shall be notified to the Secretariat by each of the Parties concerned, stating the terms of such transfer and the period for which it is to apply.

5 *bis*. Any Party not operating under paragraph 1 of Article 5 may, for one or more control periods, transfer to another such Party any portion of its calculated level of consumption set out in Article 2F, provided that the calculated level of consumption of controlled substances in Group I of Annex A of the Party transferring the portion of its calculated level of consumption did not exceed 0.25 kilograms per capita in 1989 and that the total combined calculated levels of consumption of the Parties concerned do not exceed the consumption limits set out in Article 2F. Such transfer of consumption shall be notified to the Secretariat by each of the Parties concerned, stating the terms of such transfer and the period for which it is to apply.

6. Any Party not operating under Article 5, that has facilities for the production of Annex A or Annex B controlled substances under construction, or contracted for, prior to 16 September 1987, and provided for in national legislation prior to 1 January 1987, may add the production from such facilities to its 1986 production of such substances for the purposes of determining its calculated level of production for 1986, provided that such facilities are completed by 31 December 1990 and that such production does not raise that Party's annual calculated level of consumption of the controlled substances above 0.5 kilograms per capita.

7. Any transfer of production pursuant to paragraph 5 or any addition of production pursuant to paragraph 6 shall be notified to the Secretariat, no later than the time of the transfer or addition.

8. (a) Any Parties which are Member States of a regional economic integration organization as defined in Article 1 (6) of the Convention may agree that they shall jointly fulfil their obligations respecting consumption under this Article and Articles 2A to 2I provided that their total combined calculated level of consumption does not exceed the levels required by this Article and Articles 2A to 2I.
   (b) The Parties to any such agreement shall inform the Secretariat of the terms of the agreement before the date of the reduction in consumption with which the agreement is concerned.
   (c) Such agreement will become operative only if all Member States of the regional economic integration organization and the organization concerned are Parties to the Protocol and have notified the Secretariat of their manner of implementation.

9. (a) Based on the assessments made pursuant to Article 6, the Parties may decide whether:
      (i) Adjustments to the ozone depleting potentials specified in Annex A, Annex B, Annex C and/or Annex E should be made and, if so, what the adjustments should be; and
      (ii) Further adjustments and reductions of production or consumption of the controlled substances should be undertaken and, if so, what the scope, amount and timing of any such adjustments and reductions should be;
   (b) Proposals for such adjustments shall be communicated to the Parties by the Secretariat at least six months before the meeting of the Parties at which they are proposed for adoption;

(c) In taking such decisions, the Parties shall make every effort to reach agreement by consensus. If all efforts at consensus have been exhausted, and no agreement reached, such decisions shall, as a last resort, be adopted by a two-thirds majority vote of the Parties present and voting representing a majority of the Parties operating under Paragraph 1 of Article 5 present and voting and a majority of the Parties not so operating present and voting;

(d) The decisions, which shall be binding on all Parties, shall forthwith be communicated to the Parties by the Depositary. Unless otherwise provided in the decisions, they shall enter into force on the expiry of six months from the date of the circulation of the communication by the Depositary.

10. Based on the assessments made pursuant to Article 6 of this Protocol and in accordance with the procedure set out in Article 9 of the Convention, the Parties may decide:

(a) whether any substances, and if so which, should be added to or removed from any annex to this Protocol, and

(b) the mechanism, scope and timing of the control measures that should apply to those substances;

11. Notwithstanding the provisions contained in this Article and Articles 2A to 2I Parties may take more stringent measures than those required by this Article and Articles 2A to 2I.

Article 2A
CFCs

1. Each Party shall ensure that for the twelve-month period commencing on the first day of the seventh month following the date of entry into force of this Protocol, and in each twelve-month period thereafter, its calculated level of consumption of the controlled substances in Group I of Annex A does not exceed its

calculated level of consumption in 1986. By the end of the same period, each Party producing one or more of these substances shall ensure that its calculated level of production of the substances does not exceed its calculated level of production in 1986, except that such level may have increased by no more than ten per cent based on the 1986 level. Such increase shall be permitted only so as to satisfy the basic domestic needs of the Parties operating under Article 5 and for the purposes of industrial rationalization between Parties.

2. Each Party shall ensure that for the period from 1 July 1991 to 31 December 1992 its calculated levels of consumption and production of the controlled substances in Group I of Annex A do not exceed 150 per cent of its calculated levels of production and consumption of those substances in 1986; with effect from 1 January 1993, the twelve-month control period for these controlled substances shall run from 1 January to 31 December each year.

3. Each Party shall ensure that for the twelve-month period commencing on 1 January 1994, and in each twelve-month period thereafter, its calculated level of consumption of the controlled substances in Group I of Annex A does not exceed, annually, twenty-five per cent of its calculated level of consumption in 1986. Each Party producing one or more of these substances shall, for the same periods, ensure that its calculated level of production of the substances does not exceed, annually, twenty-five per cent of its calculated level of production in 1986. However, in order to satisfy the basic domestic needs of the Parties operating under paragraph 1 of Article 5, its calculated level of production may exceed that limit by up to ten per cent of its calculated level of production in 1986.

4. Each Party shall ensure that for the twelve-month period commencing on 1 January 1996, and in each twelve-month period

thereafter, its calculated level of consumption of the controlled substances in Group I of Annex A does not exceed zero. Each Party producing one or more of these substances shall, for the same periods, ensure that its calculated level of production of the substances does not exceed zero. However, in order to satisfy the basic domestic needs of the Parties operating under paragraph 1 of Article 5, its calculated level of production may exceed that limit by a quantity equal to the annual average of its production of the controlled substances in Group I of Annex A for basic domestic needs for the period 1995 to 1997 inclusive. This paragraph will apply save to the extent that the Parties decide to permit the level of production or consumption that is necessary to satisfy uses agreed by them to be essential.

5. Each Party shall ensure that for the twelve-month period commencing on 1 January 2003 and in each twelve-month period thereafter, its calculated level of production of the controlled substances in Group I of Annex A for the basic domestic needs of the Parties operating under paragraph 1 of Article 5 does not exceed eighty per cent of the annual average of its production of those substances for basic domestic needs for the period 1995 to 1997 inclusive.

6. Each Party shall ensure that for the twelve-month period commencing on 1 January 2005 and in each twelve-month period thereafter, its calculated level of production of the controlled substances in Group I of Annex A for the basic domestic needs of the Parties operating under paragraph 1 of Article 5 does not exceed fifty per cent of the annual average of its production of those substances for basic domestic needs for the period 1995 to 1997 inclusive.

7. Each Party shall ensure that for the twelve-month period commencing on 1 January 2007 and in each twelve-month period thereafter, its calculated level of production of the controlled

substances in Group I of Annex A for the basic domestic needs of the Parties operating under paragraph 1 of Article 5 does not exceed fifteen per cent of the annual average of its production of those substances for basic domestic needs for the period 1995 to 1997 inclusive.

8. Each Party shall ensure that for the twelve-month period commencing on 1 January 2010 and in each twelve-month period thereafter, its calculated level of production of the controlled substances in Group I of Annex A for the basic domestic needs of the Parties operating under paragraph 1 of Article 5 does not exceed zero.

9. For the purposes of calculating basic domestic needs under paragraphs 4 to 8 of this Article, the calculation of the annual average of production by a Party includes any production entitlements that it has transferred in accordance with paragraph 5 of Article 2, and excludes any production entitlements that it has acquired in accordance with paragraph 5 of Article 2.

Article 2B
Halons

1. Each Party shall ensure that for the twelve-month period commencing on 1 January 1992, and in each twelve-month period thereafter, its calculated level of consumption of the controlled substances in Group II of Annex A does not exceed, annually, its calculated level of consumption in 1986. Each Party producing one or more of these substances shall, for the same periods, ensure that its calculated level of production of the substances does not exceed, annually, its calculated level of production in 1986. However, in order to satisfy the basic domestic needs of the Parties operating under paragraph 1 of Article 5, its calculated level of production may exceed that limit by up to ten per cent of its calculated level of production in 1986.

2. Each Party shall ensure that for the twelve-month period commencing on 1 January 1994, and in each twelve-month period thereafter, its calculated level of consumption of the controlled substances in Group II of Annex A does not exceed zero. Each Party producing one or more of these substances shall, for the same periods, ensure that its calculated level of production of the substances does not exceed zero. However, in order to satisfy the basic domestic needs of the Parties operating under paragraph 1 of Article 5, its calculated level of production may, until 1 January 2002 exceed that limit by up to fifteen per cent of its calculated level of production in 1986; thereafter, it may exceed that limit by a quantity equal to the annual average of its production of the controlled substances in Group II of Annex A for basic domestic needs for the period 1995 to 1997 inclusive. This paragraph will apply save to the extent that the Parties decide to permit the level of production or consumption that is necessary to satisfy uses agreed by them to be essential.

3. Each Party shall ensure that for the twelve-month period commencing on 1 January 2005 and in each twelve-month period thereafter, its calculated level of production of the controlled substances in Group II of Annex A for the basic domestic needs of the Parties operating under paragraph 1 of Article 5 does not exceed fifty per cent of the annual average of its production of those substances for basic domestic needs for the period 1995 to 1997 inclusive.

4. Each Party shall ensure that for the twelve-month period commencing on 1 January 2010 and in each twelve-month period thereafter, its calculated level of production of the controlled substances in Group II of Annex A for the basic domestic needs of the Parties operating under paragraph 1 of Article 5 does not exceed zero.

Article 2C
Other fully halogenated CFCs

1. Each Party shall ensure that for the twelve-month period commencing on 1 January 1993, its calculated level of consumption of the controlled substances in Group I of Annex B does not exceed, annually, eighty per cent of its calculated level of consumption in 1989. Each Party producing one or more of these substances shall, for the same period, ensure that its calculated level of production of the substances does not exceed, annually, eighty per cent of its calculated level of production in 1989. However, in order to satisfy the basic domestic needs of the Parties operating under paragraph 1 of Article 5, its calculated level of production may exceed that limit by up to ten per cent of its calculated level of production in 1989.

2. Each Party shall ensure that for the twelve-month period commencing on 1 January 1994, and in each twelve-month period thereafter, its calculated level of consumption of the controlled substances in Group I of Annex B does not exceed, annually, twenty-five per cent of its calculated level of consumption in 1989. Each Party producing one or more of these substances shall, for the same periods, ensure that its calculated level of production of the substances does not exceed, annually, twenty-five per cent of its calculated level of production in 1989. However, in order to satisfy the basic domestic needs of the Parties operating under paragraph 1 of Article 5, its calculated level of production may exceed that limit by up to ten per cent of its calculated level of production in 1989.

3. Each Party shall ensure that for the twelve-month period commencing on 1 January 1996, and in each twelve-month period thereafter, its calculated level of consumption of the controlled substances in Group I of Annex B does not exceed zero. Each Party producing one or more of these substances shall, for the

same periods, ensure that its calculated level of production of the substances does not exceed zero. However, in order to satisfy the basic domestic needs of the Parties operating under paragraph 1 of Article 5, its calculated level of production may, until 1 January 2003 exceed that limit by up to fifteen per cent of its calculated level of production in 1989; thereafter, it may exceed that limit by a quantity equal to eighty per cent of the annual average of its production of the controlled substances in Group I of Annex B for basic domestic needs for the period 1998 to 2000 inclusive. This paragraph will apply save to the extent that the Parties decide to permit the level of production or consumption that is necessary to satisfy uses agreed by them to be essential.

4. Each Party shall ensure that for the twelve-month period commencing on 1 January 2007 and in each twelve-month period thereafter, its calculated level of production of the controlled substances in Group I of Annex B for the basic domestic needs of the Parties operating under paragraph 1 of Article 5 does not exceed fifteen per cent of the annual average of its production of those substances for basic domestic needs for the period 1998 to 2000 inclusive.

5. Each Party shall ensure that for the twelve-month period commencing on 1 January 2010 and in each twelve-month period thereafter, its calculated level of production of the controlled substances in Group I of Annex B for the basic domestic needs of the Parties operating under paragraph 1 of Article 5 does not exceed zero.

Article 2D
Carbon tetrachloride

1. Each Party shall ensure that for the twelve-month period commencing on 1 January 1995, its calculated level of consumption of the controlled substance in Group II of Annex B

does not exceed, annually, fifteen per cent of its calculated level of consumption in 1989. Each Party producing the substance shall, for the same period, ensure that its calculated level of production of the substance does not exceed, annually, fifteen per cent of its calculated level of production in 1989. However, in order to satisfy the basic domestic needs of the Parties operating under paragraph 1 of Article 5, its calculated level of production may exceed that limit by up to ten per cent of its calculated level of production in 1989.

2. Each Party shall ensure that for the twelve-month period commencing on 1 January 1996, and in each twelve-month period thereafter, its calculated level of consumption of the controlled substance in Group II of Annex B does not exceed zero. Each Party producing the substance shall, for the same periods, ensure that its calculated level of production of the substance does not exceed zero. However, in order to satisfy the basic domestic needs of the Parties operating under paragraph 1 of Article 5, its calculated level of production may exceed that limit by up to fifteen per cent of its calculated level of production in 1989. This paragraph will apply save to the extent that the Parties decide to permit the level of production or consumption that is necessary to satisfy uses agreed by them to be essential.

Article 2E
1,1,1-Trichloroethane (Methyl chloroform)

1. Each Party shall ensure that for the twelve-month period commencing on 1 January 1993, its calculated level of consumption of the controlled substance in Group III of Annex B does not exceed, annually, its calculated level of consumption in 1989. Each Party producing the substance shall, for the same period, ensure that its calculated level of production of the substance does not exceed, annually, its calculated level of pro- duction in 1989.

However, in order to satisfy the basic domestic needs of the Parties operating under paragraph 1 of Article 5, its calculated level of production may exceed that limit by up to ten per cent of its calculated level of production in 1989.

2. Each Party shall ensure that for the twelve-month period commencing on 1 January 1994, and in each twelve-month period thereafter, its calculated level of consumption of the controlled substance in Group III of Annex B does not exceed, annually, fifty per cent of its calculated level of consumption in 1989. Each Party producing the substance shall, for the same periods, ensure that its calculated level of production of the substance does not exceed, annually, fifty per cent of its calculated level of production in 1989. However, in order to satisfy the basic domestic needs of the Parties operating under paragraph 1 of Article 5, its calculated level of production may exceed that limit by up to ten per cent of its calculated level of production in 1989.

3. Each Party shall ensure that for the twelve-month period commencing on 1 January 1996, and in each twelve-month period thereafter, its calculated level of consumption of the controlled substance in Group III of Annex B does not exceed zero. Each Party producing the substance shall, for the same periods, ensure that its calculated level of production of the substance does not exceed zero. However, in order to satisfy the basic domestic needs of the Parties operating under paragraph 1 of Article 5, its calculated level of production may exceed that limit by up to fifteen per cent of its calculated level of production for 1989. This paragraph will apply save to the extent that the Parties decide to permit the level of production or consumption that is necessary to satisfy uses agreed by them to be essential.

Article 2F
Hydrochlorofluorocarbons

1. Each Party shall ensure that for the twelve-month period commencing on 1 January 1996, and in each twelve-month period thereafter, its calculated level of consumption of the controlled substances in Group I of Annex C does not exceed, annually, the sum of:

(a) Two point eight per cent of its calculated level of consumption in 1989 of the controlled substances in Group I of Annex A; and
(b) Its calculated level of consumption in 1989 of the controlled substances in Group I of Annex C.

2. Each Party producing one or more of these substances shall ensure that for the twelve-month period commencing on 1 January 2004, and in each twelve-month period thereafter, its calculated level of production of the controlled substances in Group I of Annex C does not exceed, annually, the average of:

(a) The sum of its calculated level of consumption in 1989 of the controlled substances in Group I of Annex C and two point eight per cent of its calculated level of consumption in 1989 of the controlled substances in Group I of Annex A; and
(b) The sum of its calculated level of production in 1989 of the controlled substances in Group I of Annex C and two point eight per cent of its calculated level of production in 1989 of the controlled substances in Group I of Annex A.

However, in order to satisfy the basic domestic needs of the Parties operating under paragraph 1 of Article 5, its calculated level of production may exceed that limit by up to fifteen per cent of its calculated level of production of the controlled substances in Group I of Annex C as defined above.

3. Each Party shall ensure that for the twelve month period commencing on 1 January 2004, and in each twelve-month period thereafter, its calculated level of consumption of the controlled substances in Group I of Annex C does not exceed, annually, sixty-five per cent of the sum referred to in paragraph 1 of this Article.

4. Each Party shall ensure that for the twelve-month period commencing on 1 January 2010, and in each twelve-month period thereafter, its calculated level of consumption of the controlled substances in Group I of Annex C does not exceed, annually, twenty-five per cent of the sum referred to in paragraph 1 of this Article. Each Party producing one or more of these substances shall, for the same periods, ensure that its calculated level of production of the controlled substances in Group I of Annex C does not exceed, annually, twenty-five per cent of the calculated level referred to in paragraph 2 of this Article. However, in order to satisfy the basic domestic needs of the Parties operating under paragraph 1 of Article 5, its calculated level of production may exceed that limit by up to ten per cent of its calculated level of production of the controlled substances in Group I of Annex C as referred to in paragraph 2.

5. Each Party shall ensure that for the twelve-month period commencing on 1 January 2015, and in each twelve-month period thereafter, its calculated level of consumption of the controlled substances in Group I of Annex C does not exceed, annually, ten per cent of the sum referred to in paragraph 1 of this Article. Each Party producing one or more of these substances shall, for the same periods, ensure that its calculated level of production of the controlled substances in Group I of Annex C does not exceed, annually, ten per cent of the calculated level referred to in paragraph 2 of this Article. However, in order to satisfy the basic domestic needs of the Parties operating under paragraph 1 of Article 5, its calculated level of production may exceed that limit

by up to ten per cent of its calculated level of production of the controlled substances in Group I of Annex C as referred to in paragraph 2.

6. Each Party shall ensure that for the twelve-month period commencing on 1 January 2020, and in each twelve-month period thereafter, its calculated level of consumption of the controlled substances in Group I of Annex C does not exceed zero. Each Party producing one or more of these substances shall, for the same periods, ensure that its calculated level of production of the controlled substances in Group I of Annex C does not exceed zero. However:

(a) Each Party may exceed that limit on consumption by up to zero point five per cent of the sum referred to in paragraph 1 of this Article in any such twelve-month period ending before 1 January 2030, provided that such consumption shall be restricted to the servicing of refrigeration and air-conditioning equipment existing on 1 January 2020;
(b) Each Party may exceed that limit on production by up to zero point five per cent of the average referred to in paragraph 2 of this Article in any such twelve-month period ending before 1 January 2030, provided that such production shall be restricted to the servicing of refrigeration and air-conditioning equipment existing on 1 January 2020.

7. As of 1 January 1996, each Party shall endeavour to ensure that:

(a) The use of controlled substances in Group I of Annex C is limited to those applications where other more environmentally suitable alternative substances or technologies are not available;
(b) The use of controlled substances in Group I of Annex C is not outside the areas of application currently met by

controlled substances in Annexes A, B and C, except in rare cases for the protection of human life or human health; and

(c) Controlled substances in Group I of Annex C are selected for use in a manner that minimizes ozone depletion, in addition to meeting other environmental, safety and economic considerations.

Article 2G
Hydrobromofluorocarbons

Each Party shall ensure that for the twelve-month period commencing on 1 January 1996, and in each twelve-month period thereafter, its calculated level of consumption of the controlled substances in Group II of Annex C does not exceed zero. Each Party producing the substances shall, for the same periods, ensure that its calculated level of production of the substances does not exceed zero. This paragraph will apply save to the extent that the Parties decide to permit the level of production or consumption that is necessary to satisfy uses agreed by them to be essential.

Article 2H
Methyl bromide

1. Each Party shall ensure that for the twelve-month period commencing on 1 January 1995, and in each twelve-month period thereafter, its calculated level of consumption of the controlled substance in Annex E does not exceed, annually, its calculated level of consumption in 1991. Each Party producing the substance shall, for the same period, ensure that its calculated level of production of the substance does not exceed, annually, its calculated level of production in 1991. However, in order to satisfy the basic domestic needs of the Parties operating under paragraph 1 of Article 5, its calculated level of production may exceed that limit by up to ten per cent of its calculated level of

production in 1991.

2. Each Party shall ensure that for the twelve-month period commencing on 1 January 1999, and in the twelve-month period thereafter, its calculated level of consumption of the controlled substance in Annex E does not exceed, annually, seventy-five per cent of its calculated level of consumption in 1991. Each Party producing the substance shall, for the same periods, ensure that its calculated level of production of the substance does not exceed, annually, seventy-five per cent of its calculated level of production in 1991. However, in order to satisfy the basic domestic needs of the Parties operating under paragraph 1 of Article 5, its calculated level of production may exceed that limit by up to ten per cent of its calculated level of production in 1991.

3. Each Party shall ensure that for the twelve-month period commencing on 1 January 2001, and in the twelve-month period thereafter, its calculated level of consumption of the controlled substance in Annex E does not exceed, annually, fifty per cent of its calculated level of consumption in 1991. Each Party producing the substance shall, for the same periods, ensure that its calculated level of production of the substance does not exceed, annually, fifty per cent of its calculated level of production in 1991. However, in order to satisfy the basic domestic needs of the Parties operating under paragraph 1 of Article 5, its calculated level of production may exceed that limit by up to ten per cent of its calculated level of production in 1991.

4. Each Party shall ensure that for the twelve-month period commencing on 1 January 2003, and in the twelve-month period thereafter, its calculated level of consumption of the controlled substance in Annex E does not exceed, annually, thirty per cent of its calculated level of consumption in 1991. Each Party producing the substance shall, for the same periods, ensure that its calculated level of production of the substance does not exceed,

annually, thirty per cent of its calculated level of production in 1991. However, in order to satisfy the basic domestic needs of the Parties operating under paragraph 1 of Article 5, its calculated level of production may exceed that limit by up to ten per cent of its calculated level of production in 1991.

5. Each Party shall ensure that for the twelve-month period commencing on 1 January 2005, and in each twelve-month period thereafter, its calculated level of consumption of the controlled substance in Annex E does not exceed zero. Each Party producing the substance shall, for the same periods, ensure that its calculated level of production of the substance does not exceed zero. However, in order to satisfy the basic domestic needs of the Parties operating under paragraph 1 of Article 5, its calculated level of production may, until 1 January 2002 exceed that limit by up to fifteen per cent of its calculated level of production in 1991; thereafter, it may exceed that limit by a quantity equal to the annual average of its production of the controlled substance in Annex E for basic domestic needs for the period 1995 to 1998 inclusive. This paragraph will apply save to the extent that the Parties decide to permit the level of production or consumption that is necessary to satisfy uses agreed by them to be critical uses.

5 *bis*. Each Party shall ensure that for the twelve-month period commencing on 1 January 2005 and in each twelve-month period thereafter, its calculated level of production of the controlled substance in Annex E for the basic domestic needs of the Parties operating under paragraph 1 of Article 5 does not exceed eighty per cent of the annual average of its production of the substance for basic domestic needs for the period 1995 to 1998 inclusive.

5 *ter*. Each Party shall ensure that for the twelve-month period commencing on 1 January 2015 and in each twelve-month period thereafter, its calculated level of production of the controlled

substance in Annex E for the basic domestic needs of the Parties operating under paragraph 1 of Article 5 does not exceed zero.

6. The calculated levels of consumption and production under this Article shall not include the amounts used by the Party for quarantine and pre-shipment applications.

Article 2 I
Bromochloromethane

Each Party shall ensure that for the twelve-month period commencing on 1 January 2002, and in each twelve-month period thereafter, its calculated level of consumption and production of the controlled substance in Group III of Annex C does not exceed zero. This paragraph will apply save to the extent that the Parties decide to permit the level of production or consumption that is necessary to satisfy uses agreed by them to be essential.

Article 3
Calculation of control levels

For the purposes of Articles 2, 2A to 2I and 5, each Party shall, for each group of substances in Annex A, Annex B, Annex C or Annex E determine its calculated levels of:

(a) Production by:
  (i) multiplying its annual production of each controlled substance by the ozone depleting potential specified in respect of it in Annex A, Annex B, Annex C or Annex E;
  (ii) adding together, for each such Group, the resulting figures;
(b) Imports and exports, respectively, by following, mutatis mutandis, the procedure set out in subparagraph (a); and
(c) Consumption by adding together its calculated levels of production and imports and subtracting its calculated level

of exports as determined in accordance with subparagraphs (a) and (b). However, beginning on 1 January 1993, any export of controlled substances to non-Parties shall not be subtracted in calculating the consumption level of the exporting Party.

Article 4
Control of trade with non-Parties

1. As of 1 January 1990, each party shall ban the import of the controlled substances in Annex A from any State not party to this Protocol.

1 *bis*. Within one year of the date of the entry into force of this paragraph, each Party shall ban the import of the controlled substances in Annex B from any State not party to this Protocol.

1 *ter*. Within one year of the date of entry into force of this paragraph, each Party shall ban the import of any controlled substances in Group II of Annex C from any State not party to this Protocol.

1 *qua*. Within one year of the date of entry into force of this paragraph, each Party shall ban the import of the controlled substance in Annex E from any State not party to this Protocol.

1 *quin*. As of 1 January 2004, each Party shall ban the import of the controlled substances in Group I of Annex C from any State not party to this Protocol.

1 *sex*. Within one year of the date of entry into force of this paragraph, each Party shall ban the import of the controlled substance in Group III of Annex C from any State not party to this Protocol.

2. As of 1 January 1993, each Party shall ban the export of any controlled substances in Annex A to any State not party to this

Protocol.

2 *bis*. Commencing one year after the date of entry into force of this paragraph, each Party shall ban the export of any controlled substances in Annex B to any State not party to this Protocol.

2 *ter*. Commencing one year after the date of entry into force of this paragraph, each Party shall ban the export of any controlled substances in Group II of Annex C to any State not party to this Protocol.

2 *qua*. Commencing one year of the date of entry into force of this paragraph, each Party shall ban the export of the controlled substance in Annex E to any State not party to this Protocol.

2 *quin*. As of 1 January 2004, each Party shall ban the export of the controlled substances in Group I of Annex C to any State not party to this Protocol.

2 *sex*. Within one year of the date of entry into force of this paragraph, each Party shall ban the export of the controlled substance in Group III of Annex C to any State not party to this Protocol.

3. By 1 January 1992, the Parties shall, following the procedures in Article 10 of the Convention, elaborate in an annex a list of products containing controlled substances in Annex A. Parties that have not objected to the annex in accordance with those procedures shall ban, within one year of the annex having become effective, the import of those products from any State not party to this Protocol.

3 *bis*. Within three years of the date of the entry into force of this paragraph, the Parties shall, following the procedures in Article 10 of the Convention, elaborate in an annex a list of products containing controlled substances in Annex B. Parties that have not objected to the annex in accordance with those

procedures shall ban, within one year of the annex having become effective, the import of those products from any State not party to this Protocol.

3 *ter*. Within three years of the date of entry into force of this paragraph, the Parties shall, following the procedures in Article 10 of the Convention, elaborate in an annex a list of products containing controlled substances in Group II of Annex C. Parties that have not objected to the annex in accordance with those procedures shall ban, within one year of the annex having become effective, the import of those products from any State not party to this Protocol.

4. By 1 January 1994, the Parties shall determine the feasibility of banning or restricting, from States not party to this Protocol, the import of products produced with, but not containing, controlled substances in Annex A. If determined feasible, the Parties shall, following the procedures in Article 10 of the Convention, elaborate in an annex a list of such products. Parties that have not objected to the annex in accordance with those procedures shall ban, within one year of the annex having become effective, the import of those products from any State not party to this Protocol.

4 *bis*. Within five years of the date of the entry into force of this paragraph, the Parties shall determine the feasibility of banning or restricting, from States not party to this Protocol, the import of products produced with, but not containing, controlled substances in Annex B. If determined feasible, the Parties shall, following the procedures in Article 10 of the Convention, elaborate in an annex a list of such products. Parties that have not objected to the annex in accordance with those procedures shall ban or restrict, within one year of the annex having become effective, the import of those products from any State not party to this Protocol.

4 *ter*. Within five years of the date of entry into force of this paragraph, the Parties shall determine the feasibility of banning or restricting, from States not party to this Protocol, the import of products produced with, but not containing, controlled substances in Group II of Annex C. If determined feasible, the Parties shall, following the procedures in Article 10 of the Convention, elaborate in an annex a list of such products. Parties that have not objected to the annex in accordance with those procedures shall ban or restrict, within one year of the annex having become effective, the import of those products from any State not party to this Protocol.

5. Each Party undertakes to the fullest practicable extent to discourage the export to any State not party to this Protocol of technology for producing and for utilizing controlled substances in Annexes A, B, C and E.

6. Each Party shall refrain from providing new subsidies, aid, credits, guarantees or insurance programmes for the export to States not party to this Protocol of products, equipment, plants or technology that would facilitate the production of controlled substances in Annexes A, B, C and E.

7. Paragraphs 5 and 6 shall not apply to products, equipment, plants or technology that improve the containment, recovery, recycling or destruction of controlled substances, promote the development of alternative substances, or otherwise contribute to the reduction of emissions of controlled substances in Annexes A, B, C and E.

8. Notwithstanding the provisions of this Article, imports and exports referred to in paragraphs 1 to 4 ter of this Article may be permitted from, or to, any State not party to this Protocol, if that State is determined, by a meeting of the Parties, to be in full compliance with Article 2, Articles 2A to 2I and this Article, and have submitted data to that effect as specified in Article 7.

9. For the purposes of this Article, the term "State not party to this Protocol" shall include, with respect to a particular controlled substance, a State or regional economic integration organization that has not agreed to be bound by the control measures in effect for that substance.

10. By 1 January 1996, the Parties shall consider whether to amend this Protocol in order to extend the measures in this Article to trade in controlled substances in Group I of Annex C and in Annex E with States not party to the Protocol.

Article 4A
Control of trade with Parties

1. Where, after the phase-out date applicable to it for a controlled substance, a Party is unable, despite having taken all practicable steps to comply with its obligation under the Protocol, to cease production of that substance for domestic consumption, other than for uses agreed by the Parties to be essential, it shall ban the export of used, recycled and reclaimed quantities of that substance, other than for the purpose of destruction.

2. Paragraph 1 of this Article shall apply without prejudice to the operation of Article 11 of the Convention and the non-compliance procedure developed under Article 8 of the Protocol.

Article 4B
Licensing

1. Each Party shall, by 1 January 2000 or within three months of the date of entry into force of this Article for it, whichever is the later, establish and implement a system for licensing the import and export of new, used, recycled and reclaimed controlled substances in Annexes A, B, C and E.

2. Notwithstanding paragraph 1 of this Article, any Party operating under paragraph 1 of Article 5 which decides it is not in a position to establish and implement a system for licensing the import and export of controlled substances in Annexes C and E, may delay taking those actions until 1 January 2005 and 1 January 2002, respectively.

3. Each Party shall, within three months of the date of introducing its licensing system, report to the Secretariat on the establishment and operation of that system.

4. The Secretariat shall periodically prepare and circulate to all Parties a list of the Parties that have reported to it on their licensing systems and shall forward this information to the Implementation Committee for consideration and appropriate recommendations to the Parties.

Article 5
Special situation of developing countries

1. Any Party that is a developing country and whose annual calculated level of consumption of the controlled substances in Annex A is less than 0.3 kilograms per capita on the date of the entry into force of the Protocol for it, or any time thereafter until 1 January 1999, shall, in order to meet its basic domestic needs, be entitled to delay for ten years its compliance with the control measures set out in Articles 2A to 2E, provided that any further amendments to the adjustments or Amendment adopted at the Second Meeting of the Parties in London, 29 June 1990, shall apply to the Parties operating under this paragraph after the review provided for in paragraph 8 of this Article has taken place and shall be based on the conclusions of that review.

1 *bis*. The Parties shall, taking into account the review referred to in paragraph 8 of this Article, the assessments made pursuant to

Article 6 and any other relevant information, decide by 1 January 1996, through the procedure set forth in paragraph 9 of Article 2:

(a) With respect to paragraphs 1 to 6 of Article 2F, what base year, initial levels, control schedules and phase-out date for consumption of the controlled substances in Group I of Annex C will apply to Parties operating under paragraph 1 of this Article;
(b) With respect to Article 2G, what phase-out date for production and consumption of the controlled substances in Group II of Annex C will apply to Parties operating under paragraph 1 of this Article; and
(c) With respect to Article 2H, what base year, initial levels and control schedules for consumption and production of the controlled substance in Annex E will apply to Parties operating under paragraph 1 of this Article.

2. However, any Party operating under paragraph 1 of this Article shall exceed neither an annual calculated level of consumption of the controlled substances in Annex A of 0.3 kilograms per capita nor an annual calculated level of consumption of controlled substances of Annex B of 0.2 kilograms per capita.

3. When implementing the control measures set out in Articles 2A to 2E, any Party operating under paragraph 1 of this Article shall be entitled to use:

(a) For controlled substances under Annex A, either the average of its annual calculated level of consumption for the period 1995 to 1997 inclusive or a calculated level of consumption of 0.3 kilograms per capita, whichever is the lower, as the basis for determining its compliance with the control measures relating to consumption.
(b) For controlled substances under Annex B, the average of its

annual calculated level of consumption for the period 1998 to 2000 inclusive or a calculated level of consumption of 0.2 kilograms per capita, whichever is the lower, as the basis for determining its compliance with the control measures relating to consumption.

(c) For controlled substances under Annex A, either the average of its annual calculated level of production for the period 1995 to 1997 inclusive or a calculated level of production of 0.3 kilograms per capita, whichever is the lower, as the basis for determining its compliance with the control measures relating to production.

(d) For controlled substances under Annex B, either the average of its annual calculated level of production for the period 1998 to 2000 inclusive or a calculated level of production of 0.2 kilograms per capita, whichever is the lower, as the basis for determining its compliance with the control measures relating to production.

4. If a Party operating under paragraph 1 of this Article, at any time before the control measures obligations in Articles 2A to 2I become applicable to it, finds itself unable to obtain an adequate supply of controlled substances, it may notify this to the Secretariat. The Secretariat shall forthwith transmit a copy of such notification to the Parties, which shall consider the matter at their next Meeting, and decide upon appropriate action to be taken.

5. Developing the capacity to fulfil the obligations of the Parties operating under paragraph 1 of this Article to comply with the control measures set out in Articles 2A to 2E and Article 2I, and any control measures in Articles 2F to 2H that are decided pursuant to paragraph 1 bis of this Article, and their implementation by those same Parties will depend upon the effective implementation of the financial co-operation as provided by Article 10 and the transfer of technology as provided by Article 10A.

6. Any Party operating under paragraph 1 of this Article may, at any time, notify the Secretariat in writing that, having taken all practicable steps it is unable to implement any or all of the obligations laid down in Articles 2A to 2E and Article 2I, or any or all obligations in Articles 2F to 2H that are decided pursuant to paragraph 1 bis of this Article, due to the inadequate implementation of Articles 10 and 10A. The Secretariat shall forthwith transmit a copy of the notification to the Parties, which shall consider the matter at their next Meeting, giving due recognition to paragraph 5 of this Article and shall decide upon appropriate action to be taken.

7. During the period between notification and the Meeting of the Parties at which the appropriate action referred to in paragraph 6 above is to be decided, or for a further period if the Meeting of the Parties so decides, the non-compliance procedures referred to in Article 8 shall not be invoked against the notifying Party.

8. A Meeting of the Parties shall review, not later than 1995, the situation of the Parties operating under paragraph 1 of this Article, including the effective implementation of financial co-operation and transfer of technology to them, and adopt such revisions that may be deemed necessary regarding the schedule of control measures applicable to those Parties.

8 *bis*. Based on the conclusions of the review referred to in paragraph 8 above:

(a) With respect to the controlled substances in Annex A, a Party operating under paragraph 1 of this Article shall, in order to meet its basic domestic needs, be entitled to delay for ten years its compliance with the control measures adopted by the Second Meeting of the Parties in London, 29 June 1990, and reference by the Protocol to Articles 2A and 2B shall be read accordingly;

(b) With respect to the controlled substances in Annex B, a

Party operating under paragraph 1 of this Article shall, in order to meet its basic domestic needs, be entitled to delay for ten years its compliance with the control measures adopted by the Second Meeting of the Parties in London, 29 June 1990, and reference by the Protocol to Articles 2C to 2E shall be read accordingly.

8 *ter*. Pursuant to paragraph 1 bis above:

(a) Each Party operating under paragraph 1 of this Article shall ensure that for the twelve-month period commencing on 1 January 2013, and in each twelve-month period thereafter, its calculated level of consumption of the controlled substances in Group I of Annex C does not exceed, annually, the average of its calculated levels of consumption in 2009 and 2010. Each Party operating under paragraph 1 of this Article shall ensure that for the twelve-month period commencing on 1 January 2013 and in each twelve-month period thereafter, its calculated level of production of the controlled substances in Group I of Annex C does not exceed, annually, the average of its calculated levels of production in 2009 and 2010;

(b) Each Party operating under paragraph 1 of this Article shall ensure that for the twelve-month period commencing on 1 January 2015, and in each twelve-month period thereafter, its calculated level of consumption of the controlled substances in Group I of Annex C does not exceed, annually, ninety per cent of the average of its calculated levels of consumption in 2009 and 2010. Each such Party producing one or more of these substances shall, for the same periods, ensure that its calculated level of production of the controlled substances in Group I of Annex C does not exceed, annually, ninety per cent of the average of its calculated levels of production in 2009 and 2010;

(c) Each Party operating under paragraph 1 of this Article shall ensure that for the twelve-month period commencing on 1 January 2020, and in each twelve-month period thereafter, its calculated level of consumption of the controlled substances in Group I of Annex C does not exceed, annually, sixty-five per cent of the average of its calculated levels of consumption in 2009 and 2010. Each such Party producing one or more of these substances shall, for the same periods, ensure that its calculated level of production of the controlled substances in Group I of Annex C does not exceed, annually, sixty-five per cent of the average of its calculated levels of production in 2009 and 2010;

(d) Each Party operating under paragraph 1 of this Article shall ensure that for the twelve-month period commencing on 1 January 2025, and in each twelve-month period thereafter, its calculated level of consumption of the controlled substances in Group I of Annex C does not exceed, annually, thirty-two point five per cent of the average of its calculated levels of consumption in 2009 and 2010. Each such Party producing one or more of these substances shall, for the same periods, ensure that its calculated level of production of the controlled substances in Group I of Annex C does not exceed, annually, thirty-two point five per cent of the average of its calculated levels of production in 2009 and 2010;

(e) Each Party operating under paragraph 1 of this Article shall ensure that for the twelve-month period commencing on 1 January 2030, and in each twelve-month period thereafter, its calculated level of consumption of the controlled substances in Group I of Annex C does not exceed zero. Each such Party producing one or more of these substances shall, for the same periods, ensure that its calculated level of production of the controlled substances in Group I of Annex C does not exceed zero. However:

(i) Each such Party may exceed that limit on consumption

in any such twelve-month period so long as the sum of its calculated levels of consumption over the ten-year period from 1 January 2030 to 1 January 2040, divided by ten, does not exceed two point five per cent of the average of its calculated levels of consumption in 2009 and 2010, and provided that such consumption shall be restricted to the servicing of refrigeration and air-conditioning equipment existing on 1 January 2030;

(ii) Each such Party may exceed that limit on production in any such twelve-month period so long as the sum of its calculated levels of production over the ten-year period from 1 January 2030 to 1 January 2040, divided by ten, does not exceed two point five per cent of the average of its calculated levels of production in 2009 and 2010, and provided that such production shall be restricted to the servicing of refrigeration and air-conditioning equipment existing on 1 January 2030.

(f) Each Party operating under paragraph 1 of this Article shall comply with Article 2G;

(g) With regard to the controlled substance contained in Annex E:

(i) As of 1 January 2002 each Party operating under paragraph 1 of this Article shall comply with the control measures set out in paragraph 1 of Article 2H and, as the basis for its compliance with these control measures, it shall use the average of its annual calculated level of consumption and production, respectively, for the period of 1995 to 1998 inclusive;

(ii) Each Party operating under paragraph 1 of this Article shall ensure that for the twelve-month period commencing on 1 January 2005, and in each twelve-month period thereafter, its calculated levels of consumption and production of the controlled substance in Annex E do not exceed, annually, eighty per cent of the average of its annual calculated levels of consumption and production, respectively, for the period of 1995 to 1998 inclusive;

(iii) Each Party operating under paragraph 1 of this Article shall ensure that for the twelve-month period commencing on 1 January 2015 and in each twelve-month period thereafter, its calculated levels of consumption and production of the controlled substance in Annex E do not exceed zero. This paragraph will apply save to the extent that the Parties decide to permit the level of production or consumption that is necessary to satisfy uses agreed by them to be critical uses;

(iv) The calculated levels of consumption and production under this subparagraph shall not include the amounts used by the Party for quarantine and pre-shipment applications.

9. Decisions of the Parties referred to in paragraph 4, 6 and 7 of this Article shall be taken according to the same procedure applied to decision-making under Article 10.

Article 6
Assessment and review of control measures

Beginning in 1990, and at least every four years thereafter, the Parties shall assess the control measures provided for in Article 2 and Articles 2A to 2I on the basis of available scientific, environmental, technical and economic information. At least one year before each assessment, the Parties shall convene appropriate panels of experts qualified in the fields mentioned and determine the composition and terms of reference of any such panels. Within one year of being convened, the panels will report their conclusions, through the Secretariat, to the Parties.

Article 7
Reporting of data

1. Each Party shall provide to the Secretariat, within three months of becoming a Party, statistical data on its production, imports and exports of each of the controlled substances in Annex A for the year 1986, or the best possible estimates of such data where actual data are not available.

2. Each Party shall provide to the Secretariat statistical data on its production, imports and exports of each of the controlled substances

- in Annex B and Groups I and II of Annex C for the year 1989;
- in Annex E, for the year 1991,

or the best possible estimates of such data where actual data are not available, not later than three months after the date when the provisions set out in the Protocol with regard to the substances in Annexes B, C and E respectively enter into force for that Party.

3. Each Party shall provide to the Secretariat statistical data on its annual production (as defined in paragraph 5 of Article 1) of each of the controlled substances listed in Annexes A, B, C and E and, separately, for each substance,

- Amounts used for feedstocks,
- Amounts destroyed by technologies approved by the Parties, and
- Imports from and exports to Parties and non-Parties respectively,

for the year during which provisions concerning the substances in Annexes A, B, C and E respectively entered into force for that Party and for each year thereafter. Each Party shall provide to the Secretariat statistical data on the annual amount of the controlled substance listed in Annex E used for quarantine and pre-shipment applications. Data shall be forwarded not later than nine months after the end of the year to which the data relate.

3 *bis*. Each Party shall provide to the Secretariat separate statistical data of its annual imports and exports of each of the controlled substances listed in Group II of Annex A and Group I of Annex C that have been recycled.

4. For Parties operating under the provisions of paragraph 8 (a) of Article 2, the requirements in paragraphs 1, 2, 3 and 3 bis of this Article in respect of statistical data on imports and exports shall be satisfied if the regional economic integration organization concerned provides data on imports and exports between the organization and States that are not members of that organization.

Article 8
Non-compliance

The Parties, at their first meeting, shall consider and approve procedures and institutional mechanisms for determining non-compliance with the provisions of this Protocol and for treatment of Parties found to be in noncompliance.

Article 9
Research, development, public awareness and exchange of information

1. The Parties shall co-operate, consistent with their national laws, regulations and practices and taking into account in particular the needs of developing countries, in promoting, directly or through competent international bodies, research, development and exchange of information on:

(a) best technologies for improving the containment, recovery, recycling, or destruction of controlled substances or otherwise reducing their emissions;
(b) possible alternatives to controlled substances, to products containing such substances, and to products manufactured with them; and
(c) costs and benefits of relevant control strategies.

2. The Parties, individually, jointly or through competent international bodies, shall co-operate in promoting public awareness of the environmental effects of the emissions of controlled substances and other substances that deplete the ozone layer.

3. Within two years of the entry into force of this Protocol and every two years thereafter, each Party shall submit to the Secretariat a summary of the activities it has conducted pursuant to this Article.

Article 10
Financial mechanism

1. The Parties shall establish a mechanism for the purposes of providing financial and technical cooperation, including the transfer of technologies, to Parties operating under paragraph 1 of Article 5 of this Protocol to enable their compliance with the control measures set out in Articles 2A to 2E and Article 2I, and any control measures in Articles 2F to 2H that are decided pursuant to paragraph 1 bis of Article 5 of the Protocol. The mechanism, contributions to which shall be additional to other financial transfers to Parties operating under that paragraph, shall meet all agreed incremental costs of such Parties in order to enable their compliance with the control measures of the Protocol. An indicative list of the categories of incremental costs shall be decided by the meeting of the Parties.

2. The mechanism established under paragraph 1 shall include a Multilateral Fund. It may also include other means of multilateral, regional and bilateral co-operation.

3. The Multilateral Fund shall:

(a) Meet, on a grant or concessional basis as appropriate, and according to criteria to be decided upon by the Parties, the agreed incremental costs;

(b) Finance clearing-house functions to:

(i) Assist Parties operating under paragraph 1 of Article 5, through country specific studies and other technical co-operation, to identify their needs for co-operation;

(ii) Facilitate technical co-operation to meet these identified needs;

(iii) Distribute, as provided for in Article 9, information and relevant materials, and hold workshops, training sessions, and other related activities, for the benefit of Parties that are developing countries; and

(iv) Facilitate and monitor other multilateral, regional and bilateral co-operation available to Parties that are developing countries;

(c) Finance the secretarial services of the Multilateral Fund and related support costs.

4. The Multilateral Fund shall operate under the authority of the Parties who shall decide on its overall policies.

5. The Parties shall establish an Executive Committee to develop and monitor the implementation of specific operational policies, guidelines and administrative arrangements, including the disbursement of resources, for the purpose of achieving the objectives of the Multilateral Fund. The Executive Committee shall discharge its tasks and responsibilities, specified in its terms of reference as agreed by the Parties, with the co-operation and assistance of the International Bank for Reconstruction and Development(World

Bank), the United Nations Environment Programme, the United Nations Development Programme or other appropriate agencies depending on their respective areas of expertise. The members of the Executive Committee, which shall be selected on the basis of a balanced representation of the Parties operating under paragraph 1 of Article 5 and of the Parties not so operating, shall be endorsed by the Parties.

6. The Multilateral Fund shall be financed by contributions from Parties not operating under paragraph 1 of Article 5 in convertible currency or, in certain circumstances, in kind and/or in national currency, on the basis of the United Nations scale of assessments. Contributions by other Parties shall be encouraged. Bilateral and, in particular cases agreed by a decision of the Parties, regional co-operation may, up to a percentage and consistent with any criteria to be specified by decision of the Parties, be considered as a contribution to the Multilateral Fund, provided that such co-operation, as a minimum:

(a) Strictly relates to compliance with the provisions of this Protocol;
(b) Provides additional resources; and
(c) Meets agreed incremental costs.

7. The Parties shall decide upon the programme budget of the Multilateral Fund for each fiscal period and upon the percentage of contributions of the individual Parties thereto.

8. Resources under the Multilateral Fund shall be disbursed with the concurrence of the beneficiary Party.

9. Decisions by the Parties under this Article shall be taken by consensus whenever possible. If all efforts at consensus have been exhausted and no agreement reached, decisions shall be adopted by a two-thirds majority vote of the Parties present and voting, representing a majority of the Parties operating under paragraph 1

of Article 5 present and voting and a majority of the Parties not so operating present and voting.

10. The financial mechanism set out in this Article is without prejudice to any future arrangements that may be developed with respect to other environmental issues.

Article 10A
Transfer of technology

Each Party shall take every practicable step, consistent with the programmes supported by the financial mechanism, to ensure:

(a) that the best available, environmentally safe substitutes and related technologies are expeditiously transferred to Parties operating under paragraph 1 of Article 5; and
(b) that the transfers referred to in subparagraph (a) occur under fair and most favourable conditions.

Article 11
Meetings of the parties

1. The Parties shall hold meetings at regular intervals. The Secretariat shall convene the first meeting of the Parties not later than one year after the date of the entry into force of this Protocol and in conjunction with a meeting of the Conference of the Parties to the Convention, if a meeting of the latter is scheduled within that period.

2. Subsequent ordinary meetings of the parties shall be held, unless the Parties otherwise decide, in conjunction with meetings of the Conference of the Parties to the Convention. Extraordinary meetings of the Parties shall be held at such other times as may be deemed necessary by a meeting of the Parties, or at the written request of any Party, provided that within six months of

such a request being communicated to them by the Secretariat, it is supported by at least one third of the Parties.

3. The Parties, at their first meeting, shall:

(a) adopt by consensus rules of procedure for their meetings;
(b) adopt by consensus the financial rules referred to in paragraph 2 of Article 13;
(c) establish the panels and determine the terms of reference referred to in Article 6;
(d) consider and approve the procedures and institutional mechanisms specified in Article 8; and
(e) begin preparation of workplans pursuant to paragraph 3 of Article 10.

[The Article 10 in question is that of the original Protocol adopted in 1987]

4. The functions of the meetings of the Parties shall be to:

(a) review the implementation of this Protocol;
(b) decide on any adjustments or reductions referred to in paragraph 9 of Article 2;
(c) decide on any addition to, insertion in or removal from any annex of substances and on related control measures in accordance with paragraph 10 of Article 2;
(d) establish, where necessary, guidelines or procedures for reporting of information as provided for in Article 7 and paragraph 3 of Article 9;
(e) review requests for technical assistance submitted pursuant to paragraph 2 of Article 10;
(f) review reports prepared by the secretariat pursuant to subparagraph (c) of Article 12;
(g) assess, in accordance with Article 6, the control measures;
(h) consider and adopt, as required, proposals for amendment of this Protocol or any annex and for any new annex;

(i) consider and adopt the budget for implementing this Protocol; and
(j) consider and undertake any additional action that may be required for the achievement of the purposes of this Protocol.

5. The United Nations, its specialized agencies and the International Atomic Energy Agency, as well as any State not party to this Protocol, may be represented at meetings of the Parties as observers. Any body or agency, whether national or international, governmental or non-governmental, qualified in fields relating to the protection of the ozone layer which has informed the secretariat of its wish to be represented at a meeting of the Parties as an observer may be admitted unless at least one third of the Parties present object. The admission and participation of observers shall be subject to the rules of procedure adopted by the Parties.

Article 12
Secretariat

For the purposes of this Protocol, the Secretariat shall:

(a) arrange for and service meetings of the Parties as provided for in Article 11;
(b) receive and make available, upon request by a Party, data provided pursuant to Article 7;
(c) prepare and distribute regularly to the Parties reports based on information received pursuant to Articles 7 and 9;
(d) notify the Parties of any request for technical assistance received pursuant to Article 10 so as to facilitate the provision of such assistance;
(e) encourage non-Parties to attend the meetings of the Parties as observers and to act in accordance with the provisions of this Protocol;
(f) provide, as appropriate, the information and requests referred

to in subparagraphs (c) and (d) to such non-party observers; and

(g) perform such other functions for the achievement of the purposes of this Protocol as may be assigned to it by the Parties.

Article 13
Financial provisions

1. The funds required for the operation of this Protocol, including those for the functioning of the Secretariat related to this Protocol, shall be charged exclusively against contributions from the Parties.

2. The Parties, at their first meeting, shall adopt by consensus financial rules for the operation of this Protocol.

Article 14
Relationship of this Protocol to the Convention

Except as otherwise provided in this Protocol, the provisions of the Convention relating to its protocols shall apply to this Protocol.

Article 15
Signature

This Protocol shall be open for signature by States and by regional economic integration organizations in Montreal on 16 September 1987, in Ottawa from 17 September 1987 to 16 January 1988, and at United Nations Headquarters in New York from 17 January 1988 to 15 September 1988.

Article 16
Entry into force

1. This Protocol shall enter into force on 1 January 1989, provided that at least eleven instruments of ratification, acceptance, approval of the Protocol or accession thereto have been deposited by States or regional economic integration organizations representing at least two-thirds of 1986 estimated global consumption of the controlled substances, and the provisions of paragraph 1 of Article 17 of the Convention have been fulfilled. In the event that these conditions have not been fulfilled by that date, the Protocol shall enter into force on the ninetieth day following the date on which the conditions have been fulfilled.

2. For the purposes of paragraph 1, any such instrument deposited by a regional economic integration organization shall not be counted as additional to those deposited by member States of such organization.

3. After the entry into force of this Protocol, any State or regional economic integration organization shall become a Party to it on the ninetieth day following the date of deposit of its instrument of ratification, acceptance, approval or accession.

Article 17
Parties joining after entry into force

Subject to Article 5, any State or regional economic integration organization which becomes a Party to this Protocol after the date of its entry into force, shall fulfil forthwith the sum of the obligations under Article 2, as well as under Articles 2A to 2I and Article 4, that apply at that date to the States and regional economic integration organizations that became Parties on the date the Protocol entered into force.

Article 18
Reservations

No reservations may be made to this Protocol.

Article 19
Withdrawal

Any Party may withdraw from this Protocol by giving written notification to the Depositary at any time after four years of assuming the obligations specified in paragraph 1 of Article 2A. Any such withdrawal shall take effect upon expiry of one year after the date of its receipt by the Depositary, or on such later date as may be specified in the notification of the withdrawal.

Article 20
Authentic texts

The original of this Protocol, of which the Arabic, Chinese, English, French, Russian and Spanish texts are equally authentic, shall be deposited with the Secretary-General of the United Nations.

IN WITNESS WHEREOF the undersigned, being duly authorized to that effect, have signed this protocol.

DONE at Montreal this sixteenth day of September, one thousand nine hundred and eighty seven.

Annexes [생략]

8. United Nations Framework Convention on Climate Change (1992)

Date: 9 May 1992
In force: 21 March 1994
Contracting Parties: Signatories 165, Parties 194
Korea: 21 March 1994 (No.1213)
Link: www.unfccc.int

The Parties to this Convention,

Acknowledging that change in the Earth's climate and its adverse effects are a common concern of humankind,

Concerned that human activities have been substantially increasing the atmospheric concentrations of greenhouse gases, that these increases enhance the natural greenhouse effect, and that this will result on average in an additional warming of the Earth's surface and atmosphere and may adversely affect natural ecosystems and humankind,

Noting that the largest share of historical and current global emissions of greenhouse gases has originated in developed countries, that per capita emissions in developing countries are still relatively low and that the share of global emissions originating in developing countries will grow to meet their social and development needs,

Aware of the role and importance in terrestrial and marine ecosystems of sinks and reservoirs of greenhouse gases,

Noting that there are many uncertainties in predictions of climate change, particularly with regard to the timing, magnitude and regional patterns thereof, Acknowledging that the global nature of climate change calls for the widest possible cooperation by all countries and their participation in an effective and appropriate international response, in accordance with their common but differentiated responsibilities and respective capabilities and their social and economic conditions,

Recalling the pertinent provisions of the Declaration of the United Nations Conference on the Human Environment, adopted at Stockholm on 16 June 1972,

Recalling also that States have, in accordance with the Charter of the United Nations and the principles of international law, the sovereign right to exploit their own resources pursuant to their own environmental and developmental policies, and the responsibility to ensure that activities within their jurisdiction or control do not cause damage to the environment of other States or of areas beyond the limits of national jurisdiction,

Reaffirming the principle of sovereignty of States in international cooperation to address climate change,

Recognizing that States should enact effective environmental legislation, that environmental standards, management objectives and priorities should reflect the environmental and developmental context to which they apply, and that standards applied by some countries may be inappropriate and of unwarranted economic and social cost to other countries, in particular developing countries,

Recalling the provisions of General Assembly resolution 44/228 of 22 December 1989 on the United Nations Conference on Environment and Development, and resolutions 43/53 of 6 December 1988, 44/207 of 22 December 1989, 45/212 of 21 December 1990 and 46/169 of 19 December 1991 on protection of global climate for present and future generations of mankind,

Recalling also the provisions of General Assembly resolution 44/206 of 22 December 1989 on the possible adverse effects of sea-level rise on islands and coastal areas, particularly low-lying coastal areas and the pertinent provisions of General Assembly resolution 44/172 of 19 December 1989 on the implementation of the Plan of Action to Combat Desertification,

Recalling further the Vienna Convention for the Protection of the Ozone Layer, 1985, and the Montreal Protocol on Substances that Deplete the Ozone Layer, 1987, as adjusted and amended on 29 June 1990,

Noting the Ministerial Declaration of the Second World Climate Conference adopted on 7 November 1990,

Conscious of the valuable analytical work being conducted by many States on climate change and of the important contributions of the World Meteorological Organization, the United Nations Environment Programme and other organs, organizations and bodies of the United Nations system, as well as other international and intergovernmental bodies, to the exchange of results of scientific research and the coordination of research,

Recognizing that steps required to understand and address climate change will be environmentally, socially and economically most effective if they are based on relevant scientific, technical and economic considerations and continually re-evaluated in the light of new findings in these areas,

Recognizing that various actions to address climate change can be justified economically in their own right and can also help in solving other environmental problems,

Recognizing also the need for developed countries to take immediate action in a flexible manner on the basis of clear priorities, as a first step towards comprehensive response strategies at the global, national and, where agreed, regional levels that take into account all greenhouse gases, with due consideration of their relative contributions to the enhancement of the greenhouse effect,

Recognizing further that low-lying and other small island countries, countries with low-lying coastal, arid and semi-arid areas or areas liable to floods, drought and desertification, and developing countries with fragile mountainous ecosystems are particularly vulnerable to the adverse effects of climate change,

Recognizing the special difficulties of those countries, especially developing countries, whose economies are particularly dependent on fossil fuel production, use and exportation, as a consequence of action taken on limiting greenhouse gas emissions,

Affirming that responses to climate change should be coordinated with social and economic development in an integrated manner with a view to avoiding adverse impacts on the latter, taking into full account the legitimate priority needs of developing countries for the achievement of sustained economic growth and the eradication of poverty,

Recognizing that all countries, especially developing countries, need access to resources required to achieve sustainable social and economic development and that, in order for developing countries to progress towards that goal, their energy consumption will need to grow taking into account the possibilities for achieving greater energy efficiency and for controlling greenhouse gas emissions in general, including through the application of new technologies on terms which make such an application economically and socially beneficial,

Determined to protect the climate system for present and future generations,

Have agreed as follows:

Article 1
Definitions

For the purposes of this Convention:

1. "Adverse effects of climate change" means changes in the physical environment or biota resulting from climate change which have significant deleterious effects on the composition, resilience or productivity of natural and managed ecosystems or on the operation of socio-economic systems or on human health and welfare.

2. "Climate change" means a change of climate which is attributed directly or indirectly to human activity that alters the composition of the global atmosphere and which is in addition to natural climate variability observed over comparable time periods.

3. "Climate system" means the totality of the atmosphere, hydrosphere, biosphere and geosphere and their interactions.

4. "Emissions" means the release of greenhouse gases and/or their precursors into the atmosphere over a specified area and period of time.

5. "Greenhouse gases" means those gaseous constituents of the atmosphere, both natural and anthropogenic, that absorb and re-emit infrared radiation.

6. "Regional economic integration organization" means an organization constituted by sovereign States of a given region which has competence in respect of matters governed by this Convention or its protocols and has been duly authorized, in accordance with its internal procedures, to sign, ratify, accept, approve or accede to the instruments concerned.

7. "Reservoir" means a component or components of the climate system where a greenhouse gas or a precursor of a greenhouse

gas is stored.

8. "Sink" means any process, activity or mechanism which removes a greenhouse gas, an aerosol or a precursor of a greenhouse gas from the atmosphere.

9. "Source" means any process or activity which releases a greenhouse gas, an aerosol or a precursor of a greenhouse gas into the atmosphere.

Article 2
Objective

The ultimate objective of this Convention and any related legal instruments that the Conference of the Parties may adopt is to achieve, in accordance with the relevant provisions of the Convention, stabilization of greenhouse gas concentrations in the atmosphere at a level that would prevent dangerous anthropogenic interference with the climate system. Such a level should be achieved within a time frame sufficient to allow ecosystems to adapt naturally to climate change, to ensure that food production is not threatened and to enable economic development to proceed in a sustainable manner.

Article 3
Principles

In their actions to achieve the objective of the Convention and to implement its provisions, the Parties shall be guided, inter alia, by the following:

1. The Parties should protect the climate system for the benefit of present and future generations of humankind, on the basis of equity and in accordance with their common but differentiated responsibilities and respective capabilities. Accordingly, the

developed country Parties should take the lead in combating climate change and the adverse effects thereof.

2. The specific needs and special circumstances of developing country Parties, especially those that are particularly vulnerable to the adverse effects of climate change, and of those Parties, especially developing country Parties, that would have to bear a disproportionate or abnormal burden under the Convention, should be given full consideration.

3. The Parties should take precautionary measures to anticipate, prevent or minimize the causes of climate change and mitigate its adverse effects. Where there are threats of serious or irreversible damage, lack of full scientific certainty should not be used as a reason for postponing such measures, taking into account that policies and measures to deal with climate change should be cost-effective so as to ensure global benefits at the lowest possible cost. To achieve this, such policies and measures should take into account different socio-economic contexts, be comprehensive, cover all relevant sources, sinks and reservoirs of greenhouse gases and adaptation, and comprise all economic sectors. Efforts to address climate change may be carried out cooperatively by interested Parties.

4. The Parties have a right to, and should, promote sustainable development. Policies and measures to protect the climate system against human-induced change should be appropriate for the specific conditions of each Party and should be integrated with national development programmes, taking into account that economic development is essential for adopting measures to address climate change.

5. The Parties should cooperate to promote a supportive and open international economic system that would lead to sustainable economic growth and development in all Parties, particularly developing country Parties, thus enabling them better to address

the problems of climate change. Measures taken to combat climate change, including unilateral ones, should not constitute a means of arbitrary or unjustifiable discrimination or a disguised restriction on international trade.

Article 4
Commitments

1. All Parties, taking into account their common but differnentiated responsibilities and their specific national and regional development priorities, objectives and circumstances, shall:

(a) Develop, periodically update, publish and make available to the Conference of the Parties, in accordance with Article 12, national inventories of anthropogenic emissions by sources and removals by sinks of all greenhouse gases not controlled by the Montreal Protocol, using comparable methodologies to be agreed upon by the Conference of the Parties;

(b) Formulate, implement, publish and regularly update national and, where appropriate, regional programmes containing measures to mitigate climate change by addressing anthropogenic emissions by sources and removals by sinks of all greenhouse gases not controlled by the Montreal Protocol, and measures to facilitate adequate adaptation to climate change;

(c) Promote and cooperate in the development, application and diffusion, including transfer, of technologies, practices and processes that control, reduce or prevent anthropogenic emissions of greenhouse gases not controlled by the Montreal Protocol in all relevant sectors, including the energy, transport, industry, agriculture, forestry and waste management sectors;

(d) Promote sustainable management, and promote and cooperate in the conservation and enhancement, as appropriate, of sinks and reservoirs of all greenhouse gases not controlled by the Montreal Protocol, including biomass, forests and oceans as well as other terrestrial, coastal and marine ecosystems;

(e) Cooperate in preparing for adaptation to the impacts of climate change; develop and elaborate appropriate and integrated plans for coastal zone management, water resources and agriculture, and for the protection and rehabilitation of areas, particularly in Africa, affected by drought and desertification, as well as floods;

(f) Take climate change considerations into account, to the extent feasible, in their relevant social, economic and environmental policies and actions, and employ appropriate methods, for example impact assessments, formulated and determined nationally, with a view to minimizing adverse effects on the economy, on public health and on the quality of the environment, of projects or measures undertaken by them to mitigate or adapt to climate change;

(g) Promote and cooperate in scientific, technological, technical, socio-economic and other research, systematic observation and development of data archives related to the climate system and intended to further the understanding and to reduce or eliminate the remaining uncertainties regarding the causes, effects, magnitude and timing of climate change and the economic and social consequences of various response strategies;

(h) Promote and cooperate in the full, open and prompt exchange of relevant scientific, technological, technical, socio-economic and legal information related to the climate system and climate change, and to the economic and social consequences of various response strategies;

(i) Promote and cooperate in education, training and public awareness related to climate change and encourage the widest participation in this process, including that of non-governmental organizations; and

(j) Communicate to the Conference of the Parties information related to implementation, in accordance with Article 12.

2. The developed country Parties and other Parties included in Annex I commit themselves specifically as provided for in the following:

(a) Each of these Parties shall adopt national[2] policies and take corresponding measures on the mitigation of climate change, by limiting its anthropogenic emissions of greenhouse gases and protecting and enhancing its greenhouse gas sinks and reservoirs. These policies and measures will demonstrate that developed countries are taking the lead in modifying longer-term trends in anthropogenic emissions consistent with the objective of the Convention, recognizing that the return by the end of the present decade to earlier levels of anthropogenic emissions of carbon dioxide and other greenhouse gases not controlled by the Montreal Protocol would contribute to such modification, and taking into account the differences in these Parties' starting points and approaches, economic structures and resource bases, the need to maintain strong and sustainable economic growth, available technologies and other individual circumstances, as well as the need for equitable and appropriate contributions by each of these Parties to the global effort regarding that objective. These Parties may implement such policies and measures jointly with other Parties and may assist other Parties in contributing to the achievement of the objective of the Convention and, in particular, that of this subparagraph;

(b) In order to promote progress to this end, each of these Parties shall communicate, within six months of the entry

2) This includes policies and measures adopted by regional economic integration organizations.

into force of the Convention for it and periodically thereafter, and in accordance with Article 12, detailed information on its policies and measures referred to in subparagraph (a) above, as well as on its resulting projected anthropogenic emissions by sources and removals by sinks of greenhouse gases not controlled by the Montreal Protocol for the period referred to in subparagraph (a), with the aim of returning individually or jointly to their 1990 levels these anthropogenic emissions of carbon dioxide and other greenhouse gases not controlled by the Montreal Protocol. This information will be reviewed by the Conference of the Parties, at its first session and periodically thereafter, in accordance with Article 7;

(c) Calculations of emissions by sources and removals by sinks of greenhouse gases for the purposes of subparagraph (b) above should take into account the best available scientific knowledge, including of the effective capacity of sinks and the respective contributions of such gases to climate change. The Conference of the Parties shall consider and agree on methodologies for these calculations at its first session and review them regularly thereafter;

(d) The Conference of the Parties shall, at its first session, review the adequacy of subparagraphs (a) and (b) above. Such review shall be carried out in the light of the best available scientific information and assessment on climate change and its impacts, as well as relevant technical, social and economic information. Based on this review, the Conference of the Parties shall take appropriate action, which may include the adoption of amendments to the commitments in subparagraphs (a) and (b) above. The Conference of the Parties, at its first session, shall also take decisions regarding criteria for joint implementation as indicated in subparagraph (a) above. A second review of subparagraphs (a) and (b) shall take place not later than 31

December 1998, and thereafter at regular intervals determined by the Conference of the Parties, until the objective of the Convention is met;

(e) Each of these Parties shall:

(i) coordinate as appropriate with other such Parties, relevant economic and administrative instruments developed to achieve the objective of the Convention; and

(ii) identify and periodically review its own policies and practices which encourage activities that lead to greater levels of anthropogenic emissions of greenhouse gases not controlled by the Montreal Protocol than would otherwise occur;

(f) The Conference of the Parties shall review, not later than 31 December 1998, available information with a view to taking decisions regarding such amendments to the lists in Annexes I and II as may be appropriate, with the approval of the Party concerned;

(g) Any Party not included in Annex I may, in its instrument of ratification, acceptance, approval or accession, or at any time thereafter, notify the Depositary that it intends to be bound by subparagraphs (a) and (b) above. The Depositary shall inform the other signatories and Parties of any such notification.

3. The developed country Parties and other developed Parties included in Annex II shall provide new and additional financial resources to meet the agreed full costs incurred by developing country Parties in complying with their obligations under Article 12, paragraph 1. They shall also provide such financial resources, including for the transfer of technology, needed by the developing country Parties to meet the agreed full incremental costs of implementing measures that are covered by paragraph 1 of this Article and that are agreed between a developing country Party and the international entity or entities referred to in Article 11,

in accordance with that Article. The implementation of these commitments shall take into account the need for adequacy and predictability in the flow of funds and the importance of appropriate burden sharing among the developed country Parties.

4. The developed country Parties and other developed Parties included in Annex II shall also assist the developing country Parties that are particularly vulnerable to the adverse effects of climate change in meeting costs of adaptation to those adverse effects.

5. The developed country Parties and other developed Parties included in Annex II shall take all practicable steps to promote, facilitate and finance, as appropriate, the transfer of, or access to, environmentally sound technologies and know-how to other Parties, particularly developing country Parties, to enable them to implement the provisions of the Convention. In this process, the developed country Parties shall support the development and enhancement of endogenous capacities and technologies of developing country Parties. Other Parties and organizations in a position to do so may also assist in facilitating the transfer of such technologies.

6. In the implementation of their commitments under paragraph 2 above, a certain degree of flexibility shall be allowed by the Conference of the Parties to the Parties included in Annex I undergoing the process of transition to a market economy, in order to enhance the ability of these Parties to address climate change, including with regard to the historical level of anthropogenic emissions of greenhouse gases not controlled by the Montreal Protocol chosen as a reference.

7. The extent to which developing country Parties will effectively implement their commitments under the Convention will depend on the effective implementation by developed country Parties of their commitments under the Convention related to financial

resources and transfer of technology and will take fully into account that economic and social development and poverty eradication are the first and overriding priorities of the developing country Parties.

8. In the implementation of the commitments in this Article, the Parties shall give full consideration to what actions are necessary under the Convention, including actions related to funding, insurance and the transfer of technology, to meet the specific needs and concerns of developing country Parties arising from the adverse effects of climate change and/or the impact of the implementation of response measures, especially on:

(a) Small island countries;
(b) Countries with low-lying coastal areas;
(c) Countries with arid and semi-arid areas, forested areas and areas liable to forest decay;
(d) Countries with areas prone to natural disasters;
(e) Countries with areas liable to drought and desertification;
(f) Countries with areas of high urban atmospheric pollution;
(g) Countries with areas with fragile ecosystems, including mountainous ecosystems;
(h) Countries whose economies are highly dependent on income generated from the production, processing and export, and/or on consumption of fossil fuels and associated energy-intensive products; and
(i) Landlocked and transit countries.

Further, the Conference of the Parties may take actions, as appropriate, with respect to this paragraph.

9. The Parties shall take full account of the specific needs and special situations of the least developed countries in their actions with regard to funding and transfer of technology.

10. The Parties shall, in accordance with Article 10, take into

consideration in the implementation of the commitments of the Convention the situation of Parties, particularly developing country Parties, with economies that are vulnerable to the adverse effects of the implementation of measures to respond to climate change. This applies notably to Parties with economies that are highly dependent on income generated from the production, processing and export, and/or consumption of fossil fuels and associated energy-intensive products and/or the use of fossil fuels for which such Parties have serious difficulties in switching to alternatives.

Article 5
Research and Systematic Observation

In carrying out their commitments under Article 4, paragraph 1 (g), the Parties shall:

(a) Support and further develop, as appropriate, international and intergovernmental programmes and networks or organizations aimed at defining, conducting, assessing and financing research, data collection and systematic observation, taking into account the need to minimize duplication of effort;

(b) Support international and intergovernmental efforts to strengthen systematic observation and national scientific and technical research capacities and capabilities, particularly in developing countries, and to promote access to, and the exchange of, data and analyses thereof obtained from areas beyond national jurisdiction; and

(c) Take into account the particular concerns and needs of developing countries and cooperate in improving their endogenous capacities and capabilities to participate in the efforts referred to in subparagraphs (a) and (b) above.

Article 6
Education, Training and Public Awareness

In carrying out their commitments under Article 4, paragraph 1 (i), the Parties shall:

(a) Promote and facilitate at the national and, as appropriate, subregional and regional levels, and in accordance with national laws and regulations, and within their respective capacities:
 (i) the development and implementation of educational and public awareness programmes on climate change and its effects;
 (ii) public access to information on climate change and its effects;
 (iii) public participation in addressing climate change and its effects and developing adequate responses; and
 (iv) training of scientific, technical and managerial personnel;
(b) Cooperate in and promote, at the international level, and, where appropriate, using existing bodies:
 (i) the development and exchange of educational and public awareness material on climate change and its effects; and
 (ii) the development and implementation of education and training programmes, including the strengthening of national institutions and the exchange or secondment of personnel to train experts in this field, in particular for developing countries.

Article 7
Conference of the Parties

1. A Conference of the Parties is hereby established.

2. The Conference of the Parties, as the supreme body of this Convention, shall keep under regular review the implementation of

the Convention and any related legal instruments that the Conference of the Parties may adopt, and shall make, within its mandate, the decisions necessary to promote the effective implementation of the Convention. To this end, it shall:

(a) Periodically examine the obligations of the Parties and the institutional arrangements under the Convention, in the light of the objective of the Convention, the experience gained in its implementation and the evolution of scientific and technological knowledge;
(b) Promote and facilitate the exchange of information on measures adopted by the Parties to address climate change and its effects, taking into account the differing circumstances, responsibilities and capabilities of the Parties and their respective commitments under the Convention;
(c) Facilitate, at the request of two or more Parties, the coordination of measures adopted by them to address climate change and its effects, taking into account the differing circumstances, responsibilities and capabilities of the Parties and their respective commitments under the Convention;
(d) Promote and guide, in accordance with the objective and provisions of the Convention, the development and periodic refinement of comparable methodologies, to be agreed on by the Conference of the Parties, inter alia, for preparing inventories of greenhouse gas emissions by sources and removals by sinks, and for evaluating the effectiveness of measures to limit the emissions and enhance the removals of these gases;
(e) Assess, on the basis of all information made available to it in accordance with the provisions of the Convention, the implementation of the Convention by the Parties, the overall effects of the measures taken pursuant to the Convention, in particular environmental, economic and social effects as well as their cumulative impacts and the extent to which

progress towards the objective of the Convention is being achieved;

(f) Consider and adopt regular reports on the implementation of the Convention and ensure their publication;

(g) Make recommendations on any matters necessary for the implementation of the Convention;

(h) Seek to mobilize financial resources in accordance with Article 4, paragraphs 3, 4 and 5, and Article 11;

(i) Establish such subsidiary bodies as are deemed necessary for the implementation of the Convention;

(j) Review reports submitted by its subsidiary bodies and provide guidance to them;

(k) Agree upon and adopt, by consensus, rules of procedure and financial rules for itself and for any subsidiary bodies;

(l) Seek and utilize, where appropriate, the services and cooperation of, and information provided by, competent international organizations and intergovernmental and non-governmental bodies; and

(m) Exercise such other functions as are required for the achievement of the objective of the Convention as well as all other functions assigned to it under the Convention.

3. The Conference of the Parties shall, at its first session, adopt its own rules of procedure as well as those of the subsidiary bodies established by the Convention, which shall include decision-making procedures for matters not already covered by decision-making procedures stipulated in the Convention. Such procedures may include specified majorities required for the adoption of particular decisions.

4. The first session of the Conference of the Parties shall be convened by the interim secretariat referred to in Article 21 and shall take place not later than one year after the date of entry into force of the Convention. Thereafter, ordinary sessions of the Conference of the Parties shall be held every year unless otherwise

decided by the Conference of the Parties.

5. Extraordinary sessions of the Conference of the Parties shall be held at such other times as may be deemed necessary by the Conference, or at the written request of any Party, provided that, within six months of the request being communicated to the Parties by the secretariat, it is supported by at least one third of the Parties.

6. The United Nations, its specialized agencies and the International Atomic Energy Agency, as well as any State member thereof or observers thereto not Party to the Convention, may be represented at sessions of the Conference of the Parties as observers. Any body or agency, whether national or international, governmental or non-governmental, which is qualified in matters covered by the Convention, and which has informed the secretariat of its wish to be represented at a session of the Conference of the Parties as an observer, may be so admitted unless at least one third of the Parties present object. The admission and participation of observers shall be subject to the rules of procedure adopted by the Conference of the Parties.

Article 8
Secretariat

1. A secretariat is hereby established.

2. The functions of the secretariat shall be:

   (a) To make arrangements for sessions of the Conference of the Parties and its subsidiary bodies established under the Convention and to provide them with services as required;
   (b) To compile and transmit reports submitted to it;
   (c) To facilitate assistance to the Parties, particularly developing country Parties, on request, in the compilation and communication of information required in accordance with the

provisions of the Convention;

(d) To prepare reports on its activities and present them to the Conference of the Parties;

(e) To ensure the necessary coordination with the secretariats of other relevant international bodies;

(f) To enter, under the overall guidance of the Conference of the Parties, into such administrative and contractual arrangements as may be required for the effective discharge of its functions; and

(g) To perform the other secretariat functions specified in the Convention and in any of its protocols and such other functions as may be determined by the Conference of the Parties.

3. The Conference of the Parties, at its first session, shall designate a permanent secretariat and make arrangements for its functioning.

Article 9
Subsidiary Body for Scientific and Technological Advice

1. A subsidiary body for scientific and technological advice is hereby established to provide the Conference of the Parties and, as appropriate, its other subsidiary bodies with timely information and advice on scientific and technological matters relating to the Convention. This body shall be open to participation by all Parties and shall be multidisciplinary. It shall comprise government representatives competent in the relevant field of expertise. It shall report regularly to the Conference of the Parties on all aspects of its work.

2. Under the guidance of the Conference of the Parties, and drawing upon existing competent international bodies, this body shall:

(a) Provide assessments of the state of scientific knowledge relating to climate change and its effects;
(b) Prepare scientific assessments on the effects of measures taken in the implementation of the Convention;
(c) Identify innovative, efficient and state-of-the-art technologies and know-how and advise on the ways and means of promoting development and/or transferring such technologies;
(d) Provide advice on scientific programmes, international cooperation in research and development related to climate change, as well as on ways and means of supporting endogenous capacity-building in developing countries; and
(e) Respond to scientific, technological and methodological questions that the Conference of the Parties and its subsidiary bodies may put to the body.

3. The functions and terms of reference of this body may be further elaborated by the Conference of the Parties.

Article 10
Subsidiary Body for Implementation

1. A subsidiary body for implementation is hereby established to assist the Conference of the Parties in the assessment and review of the effective implementation of the Convention. This body shall be open to participation by all Parties and comprise government representatives who are experts on matters related to climate change. It shall report regularly to the Conference of the Parties on all aspects of its work.

2. Under the guidance of the Conference of the Parties, this body shall:

(a) Consider the information communicated in accordance with Article 12, paragraph 1, to assess the overall aggregated effect of the steps taken by the Parties in the light of the latest scientific assessments concerning climate change;

(b) Consider the information communicated in accordance with Article 12, paragraph 2, in order to assist the Conference of the Parties in carrying out the reviews required by Article 4, paragraph 2 (d); and

(c) Assist the Conference of the Parties, as appropriate, in the preparation and implementation of its decisions.

Article 11
Financial Mechanism

1. A mechanism for the provision of financial resources on a grant or concessional basis, including for the transfer of technology, is hereby defined. It shall function under the guidance of and be accountable to the Conference of the Parties, which shall decide on its policies, programme priorities and eligibility criteria related to this Convention. Its operation shall be entrusted to one or more existing international entities.

2. The financial mechanism shall have an equitable and balanced representation of all Parties within a transparent system of governance.

3. The Conference of the Parties and the entity or entities entrusted with the operation of the financial mechanism shall agree upon arrangements to give effect to the above paragraphs, which shall include the following:

(a) Modalities to ensure that the funded projects to address climate change are in conformity with the policies, programme priorities and eligibility criteria established by the Conference of the Parties;

(b) Modalities by which a particular funding decision may be reconsidered in light of these policies, programme priorities and eligibility criteria;

(c) Provision by the entity or entities of regular reports to the Conference of the Parties on its funding operations, which is consistent with the requirement for accountability set out in paragraph 1 above; and

(d) Determination in a predictable and identifiable manner of the amount of funding necessary and available for the implementation of this Convention and the conditions under which that amount shall be periodically reviewed.

4. The Conference of the Parties shall make arrangements to implement the above-mentioned provisions at its first session, reviewing and taking into account the interim arrangements referred to in Article 21, paragraph 3, and shall decide whether these interim arrangements shall be maintained. Within four years thereafter, the Conference of the Parties shall review the financial mechanism and take appropriate measures.

5. The developed country Parties may also provide and developing country Parties avail themselves of, financial resources related to the implementation of the Convention through bilateral, regional and other multilateral channels.

Article 12
Communication of Information related to Implementation

1. In accordance with Article 4, paragraph 1, each Party shall communicate to the Conference of the Parties, through the secretariat, the following elements of information:

(a) A national inventory of anthropogenic emissions by sources and removals by sinks of all greenhouse gases not controlled by the Montreal Protocol, to the extent its capacities permit, using comparable methodologies to be promoted and agreed upon by the Conference of the Parties;

(b) A general description of steps taken or envisaged by the

Party to implement the Convention; and

(c) Any other information that the Party considers relevant to the achievement of the objective of the Convention and suitable for inclusion in its communication, including, if feasible, material relevant for calculations of global emission trends.

2. Each developed country Party and each other Party included in Annex I shall incorporate in its communication the following elements of information:

(a) A detailed description of the policies and measures that it has adopted to implement its commitment under Article 4, paragraphs 2 (a) and 2 (b); and
(b) A specific estimate of the effects that the policies and measures referred to in subparagraph (a) immediately above will have on anthropogenic emissions by its sources and removals by its sinks of greenhouse gases during the period referred to in Article 4, paragraph 2 (a).

3. In addition, each developed country Party and each other developed Party included in Annex II shall incorporate details of measures taken in accordance with Article 4, paragraphs 3, 4 and 5.

4. Developing country Parties may, on a voluntary basis, propose projects for financing, including specific technologies, materials, equipment, techniques or practices that would be needed to implement such projects, along with, if possible, an estimate of all incremental costs, of the reductions of emissions and increments of removals of greenhouse gases, as well as an estimate of the consequent benefits.

5. Each developed country Party and each other Party included in Annex I shall make its initial communication within six months of the entry into force of the Convention for that Party. Each Party not so listed shall make its initial communication within

three years of the entry into force of the Convention for that Party, or of the availability of financial resources in accordance with Article 4, paragraph 3. Parties that are least developed countries may make their initial communication at their discretion. The frequency of subsequent communications by all Parties shall be determined by the Conference of the Parties, taking into account the differentiated timetable set by this paragraph.

6. Information communicated by Parties under this Article shall be transmitted by the secretariat as soon as possible to the Conference of the Parties and to any subsidiary bodies concerned. If necessary, the procedures for the communication of information may be further considered by the Conference of the Parties.

7. From its first session, the Conference of the Parties shall arrange for the provision to developing country Parties of technical and financial support, on request, in compiling and communicating information under this Article, as well as in identifying the technical and financial needs associated with proposed projects and response measures under Article 4. Such support may be provided by other Parties, by competent international organizations and by the secretariat, as appropriate.

8. Any group of Parties may, subject to guidelines adopted by the Conference of the Parties, and to prior notification to the Conference of the Parties, make a joint communication in fulfilment of their obligations under this Article, provided that such a communication includes information on the fulfilment by each of these Parties of its individual obligations under the Convention.

9. Information received by the secretariat that is designated by a Party as confidential, in accordance with criteria to be established by the Conference of the Parties, shall be aggregated by the secretariat to protect its confidentiality before being made available to any of the bodies involved in the communication and review of

information.

10. Subject to paragraph 9 above, and without prejudice to the ability of any Party to make public its communication at any time, the secretariat shall make communications by Parties under this Article publicly available at the time they are submitted to the Conference of the Parties.

Article 13
Resolution of Questions regarding Implementation

The Conference of the Parties shall, at its first session, consider the establishment of a multilateral consultative process, available to Parties on their request, for the resolution of questions regarding the implementation of the Convention.

Article 14
Settlement of Disputes

1. In the event of a dispute between any two or more Parties concerning the interpretation or application of the Convention, the Parties concerned shall seek a settlement of the dispute through negotiation or any other peaceful means of their own choice.

2. When ratifying, accepting, approving or acceding to the Convention, or at any time thereafter, a Party which is not a regional economic integration organization may declare in a written instrument submitted to the Depositary that, in respect of any dispute concerning the interpretation or application of the Convention, it recognizes as compulsory ipso facto and without special agreement, in relation to any Party accepting the same obligation:

(a) Submission of the dispute to the International Court of Justice; and/or

(b) Arbitration in accordance with procedures to be adopted by the Conference of the Parties as soon as practicable, in an annex on arbitration.

A Party which is a regional economic integration organization may make a declaration with like effect in relation to arbitration in accordance with the procedures referred to in subparagraph (b) above.

3. A declaration made under paragraph 2 above shall remain in force until it expires in accordance with its terms or until three months after written notice of its revocation has been deposited with the Depositary.

4. A new declaration, a notice of revocation or the expiry of a declaration shall not in any way affect proceedings pending before the International Court of Justice or the arbitral tribunal, unless the parties to the dispute otherwise agree.

5. Subject to the operation of paragraph 2 above, if after twelve months following notification by one Party to another that a dispute exists between them, the Parties concerned have not been able to settle their dispute through the means mentioned in paragraph 1 above, the dispute shall be submitted, at the request of any of the parties to the dispute, to conciliation.

6. A conciliation commission shall be created upon the request of one of the parties to the dispute. The commission shall be composed of an equal number of members appointed by each party concerned and a chairman chosen jointly by the members appointed by each party. The commission shall render a recommendatory award, which the parties shall consider in good faith.

7. Additional procedures relating to conciliation shall be adopted by the Conference of the Parties, as soon as practicable, in an annex on conciliation.

8. The provisions of this Article shall apply to any related legal instrument which the Conference of the Parties may adopt, unless the instrument provides otherwise.

Article 15
Amendments to the Convention

1. Any Party may propose amendments to the Convention.

2. Amendments to the Convention shall be adopted at an ordinary session of the Conference of the Parties. The text of any proposed amendment to the Convention shall be communicated to the Parties by the secretariat at least six months before the meeting at which it is proposed for adoption. The secretariat shall also communicate proposed amendments to the signatories to the Convention and, for information, to the Depositary.

3. The Parties shall make every effort to reach agreement on any proposed amendment to the Convention by consensus. If all efforts at consensus have been exhausted, and no agreement reached, the amendment shall as a last resort be adopted by a three-fourths majority vote of the Parties present and voting at the meeting. The adopted amendment shall be communicated by the secretariat to the Depositary, who shall circulate it to all Parties for their acceptance.

4. Instruments of acceptance in respect of an amendment shall be deposited with the Depositary. An amendment adopted in accordance with paragraph 3 above shall enter into force for those Parties having accepted it on the ninetieth day after the date of receipt by the Depositary of an instrument of acceptance by at least three fourths of the Parties to the Convention.

5. The amendment shall enter into force for any other Party on the ninetieth day after the date on which that Party deposits with the Depositary its instrument of acceptance of the said

amendment.

6. For the purposes of this Article, "Parties present and voting" means Parties present and casting an affirmative or negative vote.

Article 16
Adoption and Amendment of Annexes to the Convention

1. Annexes to the Convention shall form an integral part thereof and, unless otherwise expressly provided, a reference to the Convention constitutes at the same time a reference to any annexes thereto. Without prejudice to the provisions of Article 14, paragraphs 2 (b) and 7, such annexes shall be restricted to lists, forms and any other material of a descriptive nature that is of a scientific, technical, procedural or administrative character.

2. Annexes to the Convention shall be proposed and adopted in accordance with the procedure set forth in Article 15, paragraphs 2, 3 and 4.

3. An annex that has been adopted in accordance with paragraph 2 above shall enter into force for all Parties to the Convention six months after the date of the communication by the Depositary to such Parties of the adoption of the annex, except for those Parties that have notified the Depositary, in writing, within that period of their non-acceptance of the annex. The annex shall enter into force for Parties which withdraw their notification of non-acceptance on the ninetieth day after the date on which withdrawal of such notification has been received by the Depositary.

4. The proposal, adoption and entry into force of amendments to annexes to the Convention shall be subject to the same procedure as that for the proposal, adoption and entry into force of annexes to the Convention in accordance with paragraphs 2 and 3 above.

5. If the adoption of an annex or an amendment to an annex involves an amendment to the Convention, that annex or amendment to an annex shall not enter into force until such time as the amendment to the Convention enters into force.

Article 17
Protocols

1. The Conference of the Parties may, at any ordinary session, adopt protocols to the Convention.

2. The text of any proposed protocol shall be communicated to the Parties by the secretariat at least six months before such a session.

3. The requirements for the entry into force of any protocol shall be established by that instrument.

4. Only Parties to the Convention may be Parties to a protocol.

5. Decisions under any protocol shall be taken only by the Parties to the protocol concerned.

Article 18
Right to Vote

1. Each Party to the Convention shall have one vote, except as provided for in paragraph 2 below.

2. Regional economic integration organizations, in matters within their competence, shall exercise their right to vote with a number of votes equal to the number of their member States that are Parties to the Convention. Such an organization shall not exercise its right to vote if any of its member States exercises its right, and vice versa.

Article 19
Depositary

The Secretary-General of the United Nations shall be the Depositary of the Convention and of protocols adopted in accordance with Article 17.

Article 20
Signature

This Convention shall be open for signature by States Members of the United Nations or of any of its specialized agencies or that are Parties to the Statute of the International Court of Justice and by regional economic integration organizations at Rio de Janeiro, during the United Nations Conference on Environment and Development, and thereafter at United Nations Headquarters in New York from 20 June 1992 to 19 June 1993.

Article 21
Interim Arrangements

1. The secretariat functions referred to in Article 8 will be carried out on an interim basis by the secretariat established by the General Assembly of the United Nations in its resolution 45/212 of 21 December 1990, until the completion of the first session of the Conference of the Parties.

2. The head of the interim secretariat referred to in paragraph 1 above will cooperate closely with the Intergovernmental Panel on Climate Change to ensure that the Panel can respond to the need for objective scientific and technical advice. Other relevant scientific bodies could also be consulted.

3. The Global Environment Facility of the United Nations Development Programme, the United Nations Environment Pro-

gramme and the International Bank for Reconstruction and Development shall be the international entity entrusted with the operation of the financial mechanism referred to in Article 11 on an interim basis. In this connection, the Global Environment Facility should be appropriately restructured and its membership made universal to enable it to fulfil the requirements of Article 11.

Article 22
Ratification, Acceptance, Approval or Accession

1. The Convention shall be subject to ratification, acceptance, approval or accession by States and by regional economic integration organizations. It shall be open for accession from the day after the date on which the Convention is closed for signature. Instruments of ratification, acceptance, approval or accession shall be deposited with the Depositary.

2. Any regional economic integration organization which becomes a Party to the Convention without any of its member States being a Party shall be bound by all the obligations under the Convention. In the case of such organizations, one or more of whose member States is a Party to the Convention, the organization and its member States shall decide on their respective responsibilities for the performance of their obligations under the Convention. In such cases, the organization and the member States shall not be entitled to exercise rights under the Convention concurrently.

3. In their instruments of ratification, acceptance, approval or accession, regional economic integration organizations shall declare the extent of their competence with respect to the matters governed by the Convention. These organizations shall also inform the Depositary, who shall in turn inform the Parties, of any substantial modification in the extent of their competence.

Article 23
Entry into Force

1. The Convention shall enter into force on the ninetieth day after the date of deposit of the fiftieth instrument of ratification, acceptance, approval or accession.

2. For each State or regional economic integration organization that ratifies, accepts or approves the Convention or accedes thereto after the deposit of the fiftieth instrument of ratification, acceptance, approval or accession, the Convention shall enter into force on the ninetieth day after the date of deposit by such State or regional economic integration organization of its instrument of ratification, acceptance, approval or accession.

3. For the purposes of paragraphs 1 and 2 above, any instrument deposited by a regional economic integration organization shall not be counted as additional to those deposited by States members of the organization.

Article 24
Reservations

No reservations may be made to the Convention.

Article 25
Withdrawal

1. At any time after three years from the date on which the Convention as entered into force for a Party, that Party may withdraw from the Convention by giving written notification to the Depositary.

2. Any such withdrawal shall take effect upon expiry of one year from the date of receipt by the Depositary of the notification of

withdrawal, or on such later date as may be specified in the notification of withdrawal.

3. Any Party that withdraws from the Convention shall be considered as also having withdrawn from any protocol to which it is a Party.

Article 26
Authentic Texts

The original of this Convention, of which the Arabic, Chinese, English, French, Russian and Spanish texts are equally authentic, shall be deposited with the Secretary-General of the United Nations.

IN WITNESS WHEREOF the undersigned, being duly authorized to that effect, have signed this Convention.

DONE at New York this ninth day of May one thousand nine hundred and ninety-two.

Annex I

Australia	Austria	Belarus[a]
Belgium	Bulgaria[a]	Canada
Croatia[a]*	Czech Republic[a]*	Denmark
European Economic Community		Estonia[a]
Finland	France	Germany
Greece	Hungary[a]	Iceland
Ireland	Italy	Japan
Latvia[a]	Liechtenstein*	Lithuaniaa
Luxembourg	Monaco*	Netherlands
New Zealand	Norway	Poland[a]
Portugal	Romania[a]	Russian Federation[a]
Slovakia[a]*	Slovenia[a]*	Spain
Sweden	Switzerland	Turkey
Ukraine[a]	United Kingdom	United States

[a] Countries that are undergoing the process of transition to a market economy.

* Publisher's note: Countries added to Annex I by an amendment that entered into force on 13 August 1998, pursuant to decision 4/CP.3 adopted at COP.3.

Annex II

Australia	Austria	Belgium
Canada	Denmark	European Economic Community
Finland	France	Germany
Greece	Iceland	Ireland
Italy	Japan	Luxembourg
Netherlands	New Zealand	Norway
Portugal	Spain	Sweden
Switzerland	United Kingdom	United States

* Publisher's note: Turkey was deleted from Annex II by an amendment that entered into force 28 June 2002, pursuant to decision 26/CP.7 adopted at COP.7.

9. Kyoto Protocol to the United Nations Framework Convention on Climate Change (1997)

Date: 11 December 1997
In force: 16 February 2005
Contracting Parties: Signatories 84, Parties 191
Korea: 16 February 2005 (No. 1706)
Link: www.unfccc.int

The Parties to this Protocol,

Being Parties to the United Nations Framework Convention on Climate Change, hereinafter referred to as "the Convention,

In pursuit of the ultimate objective of the Convention as stated in its Article 2,

Recalling the provisions of the Convention,

Being guided by Article 3 of the Convention,

Pursuant to the Berlin Mandate adopted by decision 1/CP.1 of the Conference of the Parties to the Convention at its first session,

Have agreed as follows:

Article 1

For the purposes of this Protocol, the definitions contained in Article 1 of the Convention shall apply. In addition:

1. "Conference of the Parties" means the Conference of the Parties to the Convention.

2. "Convention" means the United Nations Framework Convention

on Climate Change, adopted in New York on 9 May 1992.

3. "Intergovernmental Panel on Climate Change" means the Intergovernmental Panel on Climate Change established in 1988 jointly by the World Meteorological Organization and the United Nations Environment Programme.

4. "Montreal Protocol" means the Montreal Protocol on Substances that Deplete the Ozone Layer, adopted in Montreal on 16 September 1987 and as subsequently adjusted and amended.

5. "Parties present and voting" means Parties present and casting an affirmative or negative vote.

6. "Party"means, unless the context otherwise indicates, a Party to this Protocol.

7. "Party included in Annex I" means a Party included in Annex I to the Convention, as may be amended, or a Party which has made a notification under Article 4, paragraph 2 (g), of the Convention.

Article 2

1. Each Party included in Annex I, in achieving its quantified emission limitation and reduction commitments under Article 3, in order to promote sustainable development, shall:

(a) Implement and/or further elaborate policies and measures in accordance with its national circumstances, such as:

(i) Enhancement of energy efficiency in relevant sectors of the national economy;

(ii) Protection and enhancement of sinks and reservoirs of greenhouse gases not controlled by the Montreal Protocol, taking into account its commitments under relevant international environmental agreements; promotion of sustainable forest management practices, afforestation and reforestation;

(iii) Promotion of sustainable forms of agriculture in light of climate change considerations;

(iv) Research on, and promotion, development and increased use of, new and renewable forms of energy, of carbon dioxide sequestration technologies and of advanced and innovative environmentally sound technologies;

(v) Progressive reduction or phasing out of market imperfections, fiscal incentives, tax and duty exemptions and subsidies in all greenhouse gas emitting sectors that run counter to the objective of the Convention and application of market instruments;

(vi) Encouragement of appropriate reforms in relevant sectors aimed at promoting policies and measures which limit or reduce emissions of greenhouse gases not controlled by the Montreal Protocol;

(vii) Measures to limit and/or reduce emissions of greenhouse gases not controlled by the Montreal Protocol in the transport sector;

(viii) Limitation and/or reduction of methane emissions through recovery and use in waste management, as well as in the production, transport and distribution of energy;

(b) Cooperate with other such Parties to enhance the individual and combined effectiveness of their policies and measures adopted under this Article, pursuant to Article 4, paragraph 2 (e) (i), of the Convention. To this end, these Parties shall take steps to share their experience and exchange information on such policies and measures, including developing ways of improving their comparability, transparency and effectiveness. The Conference of the Parties serving as the meeting of the Parties to this Protocol shall, at its first session or as soon as practicable thereafter, consider ways to facilitate such cooperation, taking into account all relevant information.

2. The Parties included in Annex I shall pursue limitation or reduction of emissions of greenhouse gases not controlled by the Montreal Protocol from aviation and marine bunker fuels, working through the International Civil Aviation Organization and the International Maritime Organization, respectively.

3. The Parties included in Annex I shall strive to implement policies and measures under this Article in such a way as to minimize adverse effects, including the adverse effects of climate change, effects on international trade, and social, environmental and economic impacts on other Parties, especially developing country Parties and in particular those identified in Article 4, paragraphs 8 and 9, of the Convention, taking into account Article 3 of the Convention. The Conference of the Parties serving as the meeting of the Parties to this Protocol may take further action, as appropriate, to promote the implementation of the provisions of this paragraph.

4. The Conference of the Parties serving as the meeting of the Parties to this Protocol, if it decides that it would be beneficial to coordinate any of the policies and measures in paragraph 1 (a) above, taking into account different national circumstances and potential effects, shall consider ways and means to elaborate the coordination of such policies and measures.

Article 3

1. The Parties included in Annex I shall, individually or jointly, ensure that their aggregate anthropogenic carbon dioxide equivalent emissions of the greenhouse gases listed in Annex A do not exceed their assigned amounts, calculated pursuant to their quantified emission limitation and reduction commitments inscribed in Annex B and in accordance with the provisions of this Article, with a view to reducing their overall emissions of such gases by at least 5 per cent below 1990 levels in the commitment period

2008 to 2012.

2. Each Party included in Annex I shall, by 2005, have made demonstrable progress in achieving its commitments under this Protocol.

3. The net changes in greenhouse gas emissions by sources and removals by sinks resulting from direct human-induced land-use change and forestry activities, limited to afforestation, reforestation and deforestation since 1990, measured as verifiable changes in carbon stocks in each commitment period, shall be used to meet the commitments under this Article of each Party included in Annex I. The greenhouse gas emissions by sources and removals by sinks associated with those activities shall be reported in a transparent and verifiable manner and reviewed in accordance with Articles 7 and 8.

4. Prior to the first session of the Conference of the Parties serving as the meeting of the Parties to this Protocol, each Party included in Annex I shall provide, for consideration by the Subsidiary Body for Scientific and Technological Advice, data to establish its level of carbon stocks in 1990 and to enable an estimate to be made of its changes in carbon stocks in subsequent years. The Conference of the Parties serving as the meeting of the Parties to this Protocol shall, at its first session or as soon as practicable thereafter, decide upon modalities, rules and guidelines as to how, and which, additional human-induced activities related to changes in greenhouse gas emissions by sources and removals by sinks in the agricultural soils and the land-use change and forestry categories shall be added to, or subtracted from, the assigned amounts for Parties included in Annex I, taking into account uncertainties, transparency in reporting, verifiability, the methodological work of the Inter-governmental Panel on Climate Change, the advice provided by the Subsidiary Body for Scientific and Technological Advice in

accordance with Article 5 and the decisions of the Conference of the Parties. Such a decision shall apply in the second and subsequent commitment periods. A Party may choose to apply such a decision on these additional human-induced activities for its first commitment period, provided that these activities have taken place since 1990.

5. The Parties included in Annex I undergoing the process of transition to a market economy whose base year or period was established pursuant to decision 9/CP.2 of the Conference of the Parties at its second session shall use that base year or period for the implementation of their commitments under this Article. Any other Party included in Annex I undergoing the process of transition to a market economy which has not yet submitted its first national communication under Article 12 of the Convention may also notify the Conference of the Parties serving as the meeting of the Parties to this Protocol that it intends to use an historical base year or period other than 1990 for the implementation of its commitments under this Article. The Conference of the Parties serving as the meeting of the Parties to this Protocol shall decide on the acceptance of such notification.

6. Taking into account Article 4, paragraph 6, of the Convention, in the implementation of their commitments under this Protocol other than those under this Article, a certain degree of flexibility shall be allowed by the Conference of the Parties serving as the meeting of the Parties to this Protocol to the Parties included in Annex I undergoing the process of transition to a market economy.

7. In the first quantified emission limitation and reduction commitment period, from 2008 to 2012, the assigned amount for each Party included in Annex I shall be equal to the percentage inscribed for it in Annex B of its aggregate anthropogenic carbon dioxide equivalent emissions of the greenhouse gases listed in

Annex A in 1990, or the base year or period determined in accordance with paragraph 5 above, multiplied by five. Those Parties included in Annex I for whom land-use change and forestry constituted a net source of greenhouse gas emissions in 1990 shall include in their 1990 emissions base year or period the aggregate anthropogenic carbon dioxide equivalent emissions by sources minus removals by sinks in 1990 from land-use change for the purposes of calculating their assigned amount.

8. Any Party included in Annex I may use 1995 as its base year for hydrofluorocarbons, perfluorocarbons and sulphur hexafluoride, for the purposes of the calculation referred to in paragraph 7 above.

9. Commitments for subsequent periods for Parties included in Annex I shall be established in amendments to Annex B to this Protocol, which shall be adopted in accordance with the provisions of Article 21, paragraph 7. The Conference of the Parties serving as the meeting of the Parties to this Protocol shall initiate the consideration of such commitments at least seven years before the end of the first commitment period referred to in paragraph 1 above.

10. Any emission reduction units, or any part of an assigned amount, which a Party acquires from another Party in accordance with the provisions of Article 6 or of Article 17 shall be added to the assigned amount for the acquiring Party.

11. Any emission reduction units, or any part of an assigned amount, which a Party transfers to another Party in accordance with the provisions of Article 6 or of Article 17 shall be subtracted from the assigned amount for the transferring Party.

12. Any certified emission reductions which a Party acquires from another Party in accordance with the provisions of Article 12 shall be added to the assigned amount for the acquiring Party.

13. If the emissions of a Party included in Annex I in a commitment period are less than its assigned amount under this Article, this difference shall, on request of that Party, be added to the assigned amount for that Party for subsequent commitment periods.

14. Each Party included in Annex I shall strive to implement the commitments mentioned in paragraph 1 above in such a way as to minimize adverse social, environmental and economic impacts on developing country Parties, particularly those identified in Article 4, paragraphs 8 and 9, of the Convention. In line with relevant decisions of the Conference of the Parties on the implementation of those paragraphs, the Conference of the Parties serving as the meeting of the Parties to this Protocol shall, at its first session, consider what actions are necessary to minimize the adverse effects of climate change and/or the impacts of response measures on Parties referred to in those paragraphs. Among the issues to be considered shall be the establishment of funding, insurance and transfer of technology.

Article 4

1. Any Parties included in Annex I that have reached an agreement to fulfil their commitments under Article 3 jointly, shall be deemed to have met those commitments provided that their total combined aggregate anthropogenic carbon dioxide equivalent emissions of the greenhouse gases listed in Annex A do not exceed their assigned amounts calculated pursuant to their quantified emission limitation and reduction commitments inscribed in Annex B and in accordance with the provisions of Article 3. The respective emission level allocated to each of the Parties to the agreement shall be set out in that agreement.

2. The Parties to any such agreement shall notify the secretariat of the terms of the agreement on the date of deposit of their

instruments of ratification, acceptance or approval of this Protocol, or accession thereto. The secretariat shall in turn inform the Parties and signatories to the Convention of the terms of the agreement.

3. Any such agreement shall remain in operation for the duration of the commitment period specified in Article 3, paragraph 7.

4. If Parties acting jointly do so in the framework of, and together with, a regional economic integration organization, any alteration in the composition of the organization after adoption of this Protocol shall not affect existing commitments under this Protocol. Any alteration in the composition of the organization shall only apply for the purposes of those commitments under Article 3 that are adopted subsequent to that alteration.

5. In the event of failure by the Parties to such an agreement to achieve their total combined level of emission reductions, each Party to that agreement shall be responsible for its own level of emissions set out in the agreement.

6. If Parties acting jointly do so in the framework of, and together with, a regional economic integration organization which is itself a Party to this Protocol, each member State of that regional economic integration organization individually, and together with the regional economic integration organization acting in accordance with Article 24, shall, in the event of failure to achieve the total combined level of emission reductions, be responsible for its level of emissions as notified in accordance with this Article.

Article 5

1. Each Party included in Annex I shall have in place, no later than one year prior to the start of the first commitment period, a national system for the estimation of anthropogenic emissions by

sources and removals by sinks of all greenhouse gases not controlled by the Montreal Protocol. Guidelines for such national systems, which shall incorporate the methodologies specified in paragraph 2 below, shall be decided upon by the Conference of the Parties serving as the meeting of the Parties to this Protocol at its first session.

2. Methodologies for estimating anthropogenic emissions by sources and removals by sinks of all greenhouse gases not controlled by the Montreal Protocol shall be those accepted by the Intergovernmental Panel on Climate Change and agreed upon by the Conference of the Parties at its third session. Where such methodologies are not used, appropriate adjustments shall be applied according to methodologies agreed upon by the Conference of the Parties serving as the meeting of the Parties to this Protocol at its first session. Based on the work of, inter alia, the Intergovernmental Panel on Climate Change and advice provided by the Subsidiary Body for Scientific and Technological Advice, the Conference of the Parties serving as the meeting of the Parties to this Protocol shall regularly review and, as appropriate, revise such methodologies and adjustments, taking fully into account any relevant decisions by the Conference of the Parties. Any revision to methodologies or adjustments shall be used only for the purposes of ascertaining compliance with commitments under Article 3 in respect of any commitment period adopted subsequent to that revision.

3. The global warming potentials used to calculate the carbon dioxide equivalence of anthropogenic emissions by sources and removals by sinks of greenhouse gases listed in Annex A shall be those accepted by the Intergovernmental Panel on Climate Change and agreed upon by the Conference of the Parties at its third session. Based on the work of, inter alia, the Intergovernmental Panel on Climate Change and advice provided by the Subsidiary Body for Scientific and Technological Advice, the Conference of

the Parties serving as the meeting of the Parties to this Protocol shall regularly review and, as appropriate, revise the global warming potential of each such greenhouse gas, taking fully into account any relevant decisions by the Conference of the Parties. Any revision to a global warming potential shall apply only to commitments under Article 3 in respect of any commitment period adopted subsequent to that revision.

Article 6

1. For the purpose of meeting its commitments under Article 3, any Party included in Annex I may transfer to, or acquire from, any other such Party emission reduction units resulting from projects aimed at reducing anthropogenic emissions by sources or enhancing anthropogenic removals by sinks of greenhouse gases in any sector of the economy, provided that:

(a) Any such project has the approval of the Parties involved;

(b) Any such project provides a reduction in emissions by sources, or an enhancement of removals by sinks, that is additional to any that would otherwise occur;

(c) It does not acquire any emission reduction units if it is not in compliance with its obligations under Articles 5 and 7; and

(d) The acquisition of emission reduction units shall be supplemental to domestic actions for the purposes of meeting commitments under Article 3.

2. The Conference of the Parties serving as the meeting of the Parties to this Protocol may, at its first session or as soon as practicable thereafter, further elaborate guidelines for the implementation of this Article, including for verification and reporting.

3. A Party included in Annex I may authorize legal entities to

participate, under its responsibility, in actions leading to the generation, transfer or acquisition under this Article of emission reduction units.

4. If a question of implementation by a Party included in Annex I of the requirements referred to in this Article is identified in accordance with the relevant provisions of Article 8, transfers and acquisitions of emission reduction units may continue to be made after the question has been identified, provided that any such units may not be used by a Party to meet its commitments under Article 3 until any issue of compliance is resolved.

Article 7

1. Each Party included in Annex I shall incorporate in its annual inventory of anthropogenic emissions by sources and removals by sinks of greenhouse gases not controlled by the Montreal Protocol, submitted in accordance with the relevant decisions of the Conference of the Parties, the necessary supplementary information for the purposes of ensuring compliance with Article 3, to be determined in accordance with paragraph 4 below.

2. Each Party included in Annex I shall incorporate in its national communication, submitted under Article 12 of the Convention, the supplementary information necessary to demonstrate compliance with its commitments under this Protocol, to be determined in accordance with paragraph 4 below.

3. Each Party included in Annex I shall submit the information required under paragraph 1 above annually, beginning with the first inventory due under the Convention for the first year of the commitment period after this Protocol has entered into force for that Party. Each such Party shall submit the information required under paragraph 2 above as part of the first national communication due under the Convention after this Protocol has

entered into force for it and after the adoption of guidelines as provided for in paragraph 4 below. The frequency of subsequent submission of information required under this Article shall be determined by the Conference of the Parties serving as the meeting of the Parties to this Protocol, taking into account any timetable for the submission of national communications decided upon by the Conference of the Parties.

4. The Conference of the Parties serving as the meeting of the Parties to this Protocol shall adopt at its first session, and review periodically thereafter, guidelines for the preparation of the information required under this Article, taking into account guidelines for the preparation of national communications by Parties included in Annex I adopted by the Conference of the Parties. The Conference of the Parties serving as the meeting of the Parties to this Protocol shall also, prior to the first commitment period, decide upon modalities for the accounting of assigned amounts.

Article 8

1. The information submitted under Article 7 by each Party included in Annex I shall be reviewed by expert review teams pursuant to the relevant decisions of the Conference of the Parties and in accordance with guidelines adopted for this purpose by the Conference of the Parties serving as the meeting of the Parties to this Protocol under paragraph 4 below. The information submitted under Article 7, paragraph 1, by each Party included in Annex I shall be reviewed as part of the annual compilation and accounting of emissions inventories and assigned amounts. Additionally, the information submitted under Article 7, paragraph 2, by each Party included in Annex I shall be reviewed as part of the review of communications.

2. Expert review teams shall be coordinated by the secretariat and

shall be composed of experts selected from those nominated by Parties to the Convention and, as appropriate, by intergovernmental organizations, in accordance with guidance provided for this purpose by the Conference of the Parties.

3. The review process shall provide a thorough and comprehensive technical assessment of all aspects of the implementation by a Party of this Protocol. The expert review teams shall prepare a report to the Conference of the Parties serving as the meeting of the Parties to this Protocol, assessing the implementation of the commitments of the Party and identifying any potential problems in, and factors influencing, the fulfilment of commitments. Such reports shall be circulated by the secretariat to all Parties to the Convention. The secretariat shall list those questions of implementation indicated in such reports for further consideration by the Conference of the Parties serving as the meeting of the Parties to this Protocol.

4. The Conference of the Parties serving as the meeting of the Parties to this Protocol shall adopt at its first session, and review periodically thereafter, guidelines for the review of implementation of this Protocol by expert review teams taking into account the relevant decisions of the Conference of the Parties.

5. The Conference of the Parties serving as the meeting of the Parties to this Protocol shall, with the assistance of the Subsidiary Body for Implementation and, as appropriate, the Subsidiary Body for Scientific and Technological Advice, consider:

(a) The information submitted by Parties under Article 7 and the reports of the expert reviews thereon conducted under this Article; and
(b) Those questions of implementation listed by the secretariat under paragraph 3 above, as well as any questions raised by Parties.

6. Pursuant to its consideration of the information referred to in paragraph above, the Conference of the Parties serving as the meeting of the Parties to this Protocol shall take decisions on any matter required for the implementation of this Protocol.

Article 9

1. The Conference of the Parties serving as the meeting of the Parties to this Protocol shall periodically review this Protocol in the light of the best available scientific information and assessments on climate change and its impacts, as well as relevant technical, social and economic information. Such reviews shall be coordinated with pertinent reviews under the Convention, in particular those required by Article 4, paragraph 2 (d), and Article 7, paragraph 2 (a), of the Convention. Based on these reviews, the Conference of the Parties serving as the meeting of the Parties to this Protocol shall take appropriate action.

2. The first review shall take place at the second session of the Conference of the Parties serving as the meeting of the Parties to this Protocol. Further reviews shall take place at regular intervals and in a timely manner.

Article 10

All Parties, taking into account their common but differentiated responsibilities and their specific national and regional development priorities, objectives and circumstances, without introducing any new commitments for Parties not included in Annex I, but reaffirming existing commitments under Article 4, paragraph 1, of the Convention, and continuing to advance the implementation of these commitments in order to achieve sustainable development, taking into account Article 4, paragraphs 3, 5 and 7, of the Convention, shall:

(a) Formulate, where relevant and to the extent possible, cost-effective national and, where appropriate, regional programmes to improve the quality of local emission factors, activity data and/or models which reflect the socio-economic conditions of each Party for the preparation and periodic updating of national inventories of anthropogenic emissions by sources and removals by sinks of all greenhouse gases not controlled by the Montreal Protocol, using comparable methodologies to be agreed upon by the Conference of the Parties, and consistent with the guidelines for the preparation of national communications adopted by the Conference of the Parties;

(b) Formulate, implement, publish and regularly update national and, where appropriate, regional programmes containing measures to mitigate climate change and measures to facilitate adequate adaptation to climate change:

(i) Such programmes would, inter alia, concern the energy, transport and industry sectors as well as agriculture, forestry and waste management. Furthermore, adaptation technologies and methods for improving spatial planning would improve adaptation to climate change; and

(ii) Parties included in Annex I shall submit information on action under this Protocol, including national programmes, in accordance with Article 7; and other Parties shall seek to include in their national communications, as appropriate, information on programmes which contain measures that the Party believes contribute to addressing climate change and its adverse impacts, including the abatement of increases in greenhouse gas emissions, and enhancement of and removals by sinks, capacity building and adaptation measures;

(c) Cooperate in the promotion of effective modalities for the development, application and diffusion of, and take all practicable steps to promote, facilitate and finance, as appropriate, the transfer of, or access to, environmentally sound technologies, know-how, practices and processes pertinent to climate change, in particular to developing countries, including the formulation of policies and programmes for the effective transfer of environmentally sound technologies that are publicly owned or in the public domain and the creation of an enabling environment for the private sector, to promote and enhance the transfer of, and access to, environmentally sound technologies;

(d) Cooperate in scientific and technical research and promote the maintenance and the development of systematic observation systems and development of data archives to reduce uncertainties related to the climate system, the adverse impacts of climate change and the economic and social consequences of various response strategies, and promote the development and strengthening of endogenous capacities and capabilities to participate in international and intergovernmental efforts, programmes and networks on research and systematic observation, taking into account Article 5 of the Convention;

(e) Cooperate in and promote at the international level, and, where appropriate, using existing bodies, the development and implementation of education and training programmes, including the strengthening of national capacity building, in particular human and institutional capacities and the exchange or secondment of personnel to train experts in this field, in particular for developing countries, and facilitate at the national level public awareness of, and public access to information on, climate change. Suitable modalities should be developed to implement these activities through the relevant bodies of the Convention, taking into account Article 6 of the Convention;

(f) Include in their national communications information on programmes and activities undertaken pursuant to this

Article in accordance with relevant decisions of the Conference of the Parties; and

(g) Give full consideration, in implementing the commitments under this Article, to Article 4, paragraph 8, of the Convention.

Article 11

1. In the implementation of Article 10, Parties shall take into account the provisions of Article 4, paragraphs 4, 5, 7, 8 and 9, of the Convention.

2. In the context of the implementation of Article 4, paragraph 1, of the Convention, in accordance with the provisions of Article 4, paragraph 3, and Article 11 of the Convention, and through the entity or entities entrusted with the operation of the financial mechanism of the Convention, the developed country Parties and other developed Parties included in Annex II to the Convention shall:

(a) Provide new and additional financial resources to meet the agreed full costs incurred by developing country Parties in advancing the implementation of existing commitments under Article 4, paragraph 1 (a), of the Convention that are covered in Article 10, subparagraph (a); and

(b) Also provide such financial resources, including for the transfer of technology, needed by the developing country Parties to meet the agreed full incremental costs of advancing the implementation of existing commitments under Article 4, paragraph 1, of the Convention that are covered by Article 10 and that are agreed between a developing country Party and the international entity or entities referred to in Article 11 of the Convention, in accordance with that Article.

The implementation of these existing commitments shall take into account the need for adequacy and predictability in the flow of funds and the importance of appropriate burden sharing among developed country Parties. The guidance to the entity or entities entrusted with the operation of the financial mechanism of the Convention in relevant decisions of the Conference of the Parties, including those agreed before the adoption of this Protocol, shall apply mutatis mutandis to the provisions of this paragraph.

3. The developed country Parties and other developed Parties in Annex II to the Convention may also provide, and developing country Parties avail themselves of, financial resources for the implementation of Article 10, through bilateral, regional and other multilateral channels.

Article 12

1. A clean development mechanism is hereby defined.

2. The purpose of the clean development mechanism shall be to assist Parties not included in Annex I in achieving sustainable development and in contributing to the ultimate objective of the Convention, and to assist Parties included in Annex I in achieving compliance with their quantified emission limitation and reduction commitments under Article 3.

3. Under the clean development mechanism:

(a) Parties not included in Annex I will benefit from project activities resulting in certified emission reductions; and

(b) Parties included in Annex I may use the certified emission reductions accruing from such project activities to contribute to compliance with part of their quantified emission limitation and reduction commitments under Article 3, as determined by the Conference of the Parties serving as the meeting of the Parties to this Protocol.

4. The clean development mechanism shall be subject to the authority and guidance of the Conference of the Parties serving as the meeting of the Parties to this Protocol and be supervised by an executive board of the clean development mechanism.

5. Emission reductions resulting from each project activity shall be certified by operational entities to be designated by the Conference of the Parties serving as the meeting of the Parties to this Protocol, on the basis of:

(a) Voluntary participation approved by each Party involved;
(b) Real, measurable, and long-term benefits related to the mitigation of climate change; and
(c) Reductions in emissions that are additional to any that would occur in the absence of the certified project activity.

6. The clean development mechanism shall assist in arranging funding of certified project activities as necessary.

7. The Conference of the Parties serving as the meeting of the Parties to this Protocol shall, at its first session, elaborate modalities and procedures with the objective of ensuring transparency, efficiency and accountability through independent auditing and verification of project activities.

8. The Conference of the Parties serving as the meeting of the Parties to this Protocol shall ensure that a share of the proceeds from certified project activities is used to cover administrative expenses as well as to assist developing country Parties that are particularly vulnerable to the adverse effects of climate change to meet the costs of adaptation.

9. Participation under the clean development mechanism, including in activities mentioned in paragraph 3 (a) above and in the acquisition of certified emission reductions, may involve private and/or public entities, and is to be subject to whatever guidance may be provided by the executive board of the clean development

mechanism.

10. Certified emission reductions obtained during the period from the year 2000 up to the beginning of the first commitment period can be used to assist in achieving compliance in the first commitment period.

Article 13

1. The Conference of the Parties, the supreme body of the Convention, shall serve as the meeting of the Parties to this Protocol.

2. Parties to the Convention that are not Parties to this Protocol may participate as observers in the proceedings of any session of the Conference of the Parties serving as the meeting of the Parties to this Protocol. When the Conference of the Parties serves as the meeting of the Parties to this Protocol, decisions under this Protocol shall be taken only by those that are Parties to this Protocol.

3. When the Conference of the Parties serves as the meeting of the Parties to this Protocol, any member of the Bureau of the Conference of the Parties representing a Party to the Convention but, at that time, not a Party to this Protocol, shall be replaced by an additional member to be elected by and from amongst the Parties to this Protocol.

4. The Conference of the Parties serving as the meeting of the Parties to this Protocol shall keep under regular review the implementation of this Protocol and shall make, within its mandate, the decisions necessary to promote its effective implementation. It shall perform the functions assigned to it by this Protocol and shall:

(a) Assess, on the basis of all information made available to it in accordance with the provisions of this Protocol, the implementation of this Protocol by the Parties, the overall effects of the measures taken pursuant to this Protocol, in particular environmental, economic and social effects as well as their cumulative impacts and the extent to which progress towards the objective of the Convention is being achieved;

(b) Periodically examine the obligations of the Parties under this Protocol, giving due consideration to any reviews required by Article 4, paragraph 2 (d), and Article 7, paragraph 2, of the Convention, in the light of the objective of the Convention, the experience gained in its implementation and the evolution of scientific and technological knowledge, and in this respect consider and adopt regular reports on the implementation of this Protocol;

(c) Promote and facilitate the exchange of information on measures adopted by the Parties to address climate change and its effects, taking into account the differing circumstances, responsibilities and capabilities of the Parties and their respective commitments under this Protocol;

(d) Facilitate, at the request of two or more Parties, the coordination of measures adopted by them to address climate change and its effects, taking into account the differing circumstances, responsibilities and capabilities of the Parties and their respective commitments under this Protocol;

(e) Promote and guide, in accordance with the objective of the Convention and the provisions of this Protocol, and taking fully into account the relevant decisions by the Conference of the Parties, the development and periodic refinement of comparable methodologies for the effective implementation of this Protocol, to be agreed on by the Conference of the Parties serving as the meeting of the Parties to this Protocol;

(f) Make recommendations on any matters necessary for the

implementation of this Protocol;

(g) Seek to mobilize additional financial resources in accordance with Article 11, paragraph 2;

(h) Establish such subsidiary bodies as are deemed necessary for the implementation of this Protocol;

(i) Seek and utilize, where appropriate, the services and cooperation of, and information provided by, competent international organizations and intergovernmental and non-governmental bodies; and

(j) Exercise such other functions as may be required for the implementation of this Protocol, and consider any assignment resulting from a decision by the Conference of the Parties.

5. The rules of procedure of the Conference of the Parties and financial procedures applied under the Convention shall be applied mutatis mutandis under this Protocol, except as may be otherwise decided by consensus by the Conference of the Parties serving as the meeting of the Parties to this Protocol.

6. The first session of the Conference of the Parties serving as the meeting of the Parties to this Protocol shall be convened by the secretariat in conjunction with the first session of the Conference of the Parties that is scheduled after the date of the entry into force of this Protocol. Subsequent ordinary sessions of the Conference of the Parties serving as the meeting of the Parties to this Protocol shall be held every year and in conjunction with ordinary sessions of the Conference of the Parties, unless otherwise decided by the Conference of the Parties serving as the meeting of the Parties to this Protocol.

7. Extraordinary sessions of the Conference of the Parties serving as the meeting of the Parties to this Protocol shall be held at such other times as may be deemed necessary by the Conference of the Parties serving as the meeting of the Parties to this Protocol, or at the written request of any Party, provided that,

within six months of the request being communicated to the Parties by the secretariat, it is supported by at least one third of the Parties.

8. The United Nations, its specialized agencies and the International Atomic Energy Agency, as well as any State member thereof or observers thereto not party to the Convention, may be represented at sessions of the Conference of the Parties serving as the meeting of the Parties to this Protocol as observers. Any body or agency, whether national or international, governmental or non-governmental, which is qualified in matters covered by this Protocol and which has informed the secretariat of its wish to be represented at a session of the Conference of the Parties serving as the meeting of the Parties to this Protocol as an observer, may be so admitted unless at least one third of the Parties present object. The admission and participation of observers shall be subject to the rules of procedure, as referred to in paragraph 5 above.

Article 14

1. The secretariat established by Article 8 of the Convention shall serve as the secretariat of this Protocol.

2. Article 8, paragraph 2, of the Convention on the functions of the secretariat, and Article 8, paragraph 3, of the Convention on arrangements made for the functioning of the secretariat, shall apply mutatis mutandis to this Protocol. The secretariat shall, in addition, exercise the functions assigned to it under this Protocol.

Article 15

1. The Subsidiary Body for Scientific and Technological Advice and the Subsidiary Body for Implementation established by Articles 9 and 10 of the Convention shall serve as, respectively,

the Subsidiary Body for Scientific and Technological Advice and the Subsidiary Body for Implementation of this Protocol. The provisions relating to the functioning of these two bodies under the Convention shall apply mutatis mutandis to this Protocol. Sessions of the meetings of the Subsidiary Body for Scientific and Technological Advice and the Subsidiary Body for Implementation of this Protocol shall be held in conjunction with the meetings of, respectively, the Subsidiary Body for Scientific and Technological Advice and the Subsidiary Body for Implementation of the Convention.

2. Parties to the Convention that are not Parties to this Protocol may participate as observers in the proceedings of any session of the subsidiary bodies. When the subsidiary bodies serve as the subsidiary bodies of this Protocol, decisions under this Protocol shall be taken only by those that are Parties to this Protocol.

3. When the subsidiary bodies established by Articles 9 and 10 of the Convention exercise their functions with regard to matters concerning this Protocol, any member of the Bureaux of those subsidiary bodies representing a Party to the Convention but, at that time, not a party to this Protocol, shall be replaced by an additional member to be elected by and from amongst the Parties to this Protocol.

Article 16

The Conference of the Parties serving as the meeting of the Parties to this Protocol shall, as soon as practicable, consider the application to this Protocol of, and modify as appropriate, the multilateral consultative process referred to in Article 13 of the Convention, in the light of any relevant decisions that may be taken by the Conference of the Parties. Any multilateral consultative process that may be applied to this Protocol shall operate without prejudice to the procedures and mechanisms established in accordance with Article 18.

Article 17

The Conference of the Parties shall define the relevant principles, modalities, rules and guidelines, in particular for verification, reporting and accountability for emissions trading. The Parties included in Annex B may participate in emissions trading for the purposes of fulfilling their commitments under Article 3. Any such trading shall be supplemental to domestic actions for the purpose of meeting quantified emission limitation and reduction commitments under that Article.

Article 18

The Conference of the Parties serving as the meeting of the Parties to this Protocol shall, at its first session, approve appropriate and effective procedures and mechanisms to determine and to address cases of non-compliance with the provisions of this Protocol, including through the development of an indicative list of consequences, taking into account the cause, type, degree and frequency of non-compliance. Any procedures and mechanisms under this Article entailing binding consequences shall be adopted by means of an amendment to this Protocol.

Article 19

The provisions of Article 14 of the Convention on settlement of disputes shall apply mutatis mutandis to this Protocol.

Article 20

1. Any Party may propose amendments to this Protocol.

2. Amendments to this Protocol shall be adopted at an ordinary session of the Conference of the Parties serving as the meeting of the Parties to this Protocol. The text of any proposed amendment to this Protocol shall be communicated to the Parties by the secretariat at least six months before the meeting at which it is proposed for adoption. The secretariat shall also communicate the text of any proposed amendments to the Parties and signatories to the Convention and, for information, to the Depositary.

3. The Parties shall make every effort to reach agreement on any proposed amendment to this Protocol by consensus. If all efforts at consensus have been exhausted, and no agreement reached, the amendment shall as a last resort be adopted by a three-fourths majority vote of the Parties present and voting at the meeting. The adopted amendment shall be communicated by the secretariat to the Depositary, who shall circulate it to all Parties for their acceptance.

4. Instruments of acceptance in respect of an amendment shall be deposited with the Depositary. An amendment adopted in accordance with paragraph 3 above shall enter into force for those Parties having accepted it on the ninetieth day after the date of receipt by the Depositary of an instrument of acceptance by at least three fourths of the Parties to this Protocol.

5. The amendment shall enter into force for any other Party on the ninetieth day after the date on which that Party deposits with the Depositary its instrument of acceptance of the said amendment.

Article 21

1. Annexes to this Protocol shall form an integral part thereof and, unless otherwise expressly provided, a reference to this Protocol constitutes at the same time a reference to any annexes

thereto. Any annexes adopted after the entry into force of this Protocol shall be restricted to lists, forms and any other material of a descriptive nature that is of a scientific, technical, procedural or administrative character.

2. Any Party may make proposals for an annex to this Protocol and may propose amendments to annexes to this Protocol.

3. Annexes to this Protocol and amendments to annexes to this Protocol shall be adopted at an ordinary session of the Conference of the Parties serving as the meeting of the Parties to this Protocol. The text of any proposed annex or amendment to an annex shall be communicated to the Parties by the secretariat at least six months before the meeting at which it is proposed for adoption. The secretariat shall also communicate the text of any proposed annex or amendment to an annex to the Parties and signatories to the Convention and, for information, to the Depositary.

4. The Parties shall make every effort to reach agreement on any proposed annex or amendment to an annex by consensus. If all efforts at consensus have been exhausted, and no agreement reached, the annex or amendment to an annex shall as a last resort be adopted by a three-fourths majority vote of the Parties present and voting at the meeting. The adopted annex or amendment to an annex shall be communicated by the secretariat to the Depositary, who shall circulate it to all Parties for their acceptance.

5. An annex, or amendment to an annex other than Annex A or B, that has been adopted in accordance with paragraphs 3 and 4 above shall enter into force for all Parties to this Protocol six months after the date of the communication by the Depositary to such Parties of the adoption of the annex or adoption of the amendment to the annex, except for those Parties that have notified the Depositary, in writing, within that period of their

non-acceptance of the annex or amendment to the annex. The annex or amendment to an annex shall enter into force for Parties which withdraw their notification of non-acceptance on the ninetieth day after the date on which withdrawal of such notification has been received by the Depositary.

6. If the adoption of an annex or an amendment to an annex involves an amendment to this Protocol, that annex or amendment to an annex shall not enter into force until such time as the amendment to this Protocol enters into force.

7. Amendments to Annexes A and B to this Protocol shall be adopted and enter into force in accordance with the procedure set out in Article 20, provided that any amendment to Annex B shall be adopted only with the written consent of the Party concerned.

Article 22

1. Each Party shall have one vote, except as provided for in paragraph 2 below.

2. Regional economic integration organizations, in matters within their competence, shall exercise their right to vote with a number of votes equal to the number of their member States that are Parties to this Protocol. Such an organization shall not exercise its right to vote if any of its member States exercises its right, and vice versa.

Article 23

The Secretary-General of the United Nations shall be the Depositary of this Protocol.

Article 24

1. This Protocol shall be open for signature and subject to ratification, acceptance or approval by States and regional economic integration organizations which are Parties to the Convention. It shall be open for signature at United Nations Headquarters in New York from 16 March 1998 to 15 March 1999. This Protocol shall be open for accession from the day after the date on which it is closed for signature. Instruments of ratification, acceptance, approval or accession shall be deposited with the Depositary.

2. Any regional economic integration organization which becomes a Party to this Protocol without any of its member States being a Party shall be bound by all the obligations under this Protocol. In the case of such organizations, one or more of whose member States is a Party to this Protocol, the organization and its member States shall decide on their respective responsibilities for the performance of their obligations under this Protocol. In such cases, the organization and the member States shall not be entitled to exercise rights under this Protocol concurrently.

3. In their instruments of ratification, acceptance, approval or accession, regional economic integration organizations shall declare the extent of their competence with respect to the matters governed by this Protocol. These organizations shall also inform the Depositary, who shall in turn inform the Parties, of any substantial modification in the extent of their competence.

Article 25

1. This Protocol shall enter into force on the ninetieth day after the date on which not less than 55 Parties to the Convention, incorporating Parties included in Annex I which accounted in total for at least 55 per cent of the total carbon dioxide

emissions for 1990 of the Parties included in Annex I, have deposited their instruments of ratification, acceptance, approval or accession.

2. For the purposes of this Article, “"the total carbon dioxide emissions for 1990 of the Parties included in Annex I”" means the amount communicated on or before the date of adoption of this Protocol by the Parties included in Annex I in their first national communications submitted in accordance with Article 12 of the Convention.

3. For each State or regional economic integration organization that ratifies, accepts or approves this Protocol or accedes thereto after the conditions set out in paragraph 1 above for entry into force have been fulfilled, this Protocol shall enter into force on the ninetieth day following the date of deposit of its instrument of ratification, acceptance, approval or accession.

4. For the purposes of this Article, any instrument deposited by a regional economic integration organization shall not be counted as additional to those deposited by States members of the organization.

Article 26

No reservations may be made to this Protocol.

Article 27

1. At any time after three years from the date on which this Protocol has entered into force for a Party, that Party may withdraw from this Protocol by giving written notification to the Depositary.

2. Any such withdrawal shall take effect upon expiry of one year from the date of receipt by the Depositary of the notification of

withdrawal, or on such later date as may be specified in the notification of withdrawal.

3. Any Party that withdraws from the Convention shall be considered as also having withdrawn from this Protocol.

Article 28

The original of this Protocol, of which the Arabic, Chinese, English, French, Russian and Spanish texts are equally authentic, shall be deposited with the Secretary-General of the United Nations.

DONE at Kyoto this eleventh day of December one thousand nine hundred and ninety-seven.

IN WITNESS WHEREOF the undersigned, being duly authorized to that effect, have affixed their signatures to this Protocol on the dates indicated.

Annex A

Greenhouse gases

Carbon dioxide(C02)
Methane(CH4)
Nitrous oxide(N20)
Hydrofluorocarbons(HFCs)
Perfluorocarbons(PFCs)
Sulphur hexafluoride(SF6)

Sectors/source categories

Energy

Fuel combustion

        - Energy industries
        - Manufacturing industries and construction
        - Transport
        - Other sectors
        - Other
    - Fugitive emissions from fuels
        - Solid fuels
        - Oil and natural gas
        - Other
- Industrial processes
    - Mineral products
    - Chemical industry
    - Metal production
    - Other production
    - Production of halocarbons and sulphur hexafluoride
    - Consumption of halocarbons and sulphur hexafluoride
    - Other
- Solvent and other product use
- Agriculture
    - Enteric fermentation
    - Manure management
    - Rice cultivation
    - Agricultural soils
    - Prescribed burning of savannas
    - Field burning of agricultural residues
    - Other
- Waste
    - Solid waste disposal on land
    - Wastewater handling
    - Waste incineration
    - Other

Annex B
Party / Quantified emission limitation or reduction commitment (percentage of base year or period)

Party	%	Party	%
Australia	108	Liechtenstein	92
Austria	92	Lithuania*	92
Belgium	92	Luxembourg	92
Bulgaria*	92	Monaco	92
Canada	94	Netherlands	92
Croatia*	95	New Zealand	100
Czech Republic*	92	Norway	101
Denmark	92	Poland*	94
Estonia*	92	Portugal	92
European Community	92	Romania*	92
Finland	92	Russian Federation*	100
France	92	Slovakia*	92
Germany	92	Slovenia*	92
Greece	92	Spain	92
Hungary*	94	Sweden	92
Iceland	110	Switzerland	92
Ireland	92	Ukraine*	100
Italy	92	United Kingdom	92
Japan	94	United States	93
Latvia*	92		

* Countries that are undergoing the process of transition to a market economy.

10. Johannesburg Declaration on Sustainable Development (2002)

A/CONF.199/20
Date: 4 September 2002
Link: www.un-documents.net

From our Origins to the Future

1. We, the representatives of the peoples of the world, assembled at the World Summit on Sustainable Development in Johannesburg, South Africa from 2-4 September 2002, reaffirm our commitment to sustainable development.

2. We commit ourselves to build a humane, equitable and caring global society cognizant of the need for human dignity for all.

3. At the beginning of this Summit, the children of the world spoke to us in a simple yet clear voice that the future belongs to them, and accordingly challenged all of us to ensure that through our actions they will inherit a world free of the indignity and indecency occasioned by poverty, environmental degradation and patterns of unsustainable development.

4. As part of our response to these children, who represent our collective future, all of us, coming from every corner of the world, informed by different life experiences, are united and moved by a deeply-felt sense that we urgently need to create a new and brighter world of hope.

5. Accordingly, we assume a collective responsibility to advance and strengthen the interdependent and mutually reinforcing pillars of sustainable development – economic development, social devel-

opment and environmental protection – at local, national, regional and global levels.

6. From this Continent, the Cradle of Humanity we declare, through the Plan of Implementation and this Declaration, our responsibility to one another, to the greater community of life and to our children.

7. Recognizing that humankind is at a crossroad, we have united in a common resolve to make a determined effort to respond positively to the need to produce a practical and visible plan that should bring about poverty eradication and human development.

From Stockholm to Rio de Janeiro to Johannesburg

8. Thirty years ago, in Stockholm, we agreed on the urgent need to respond to the problem of environmental deterioration. Ten years ago, at the United Nations Conference on Environment and Development, held in Rio de Janeiro, we agreed that the protection of the environment, and social and economic development are fundamental to sustainable development, based on the Rio Principles. To achieve such development, we adopted the global programme, Agenda 21, and the Rio Declaration, to which we reaffirm our commitment. The Rio Summit was a significant milestone that set a new agenda for sustainable development.

9. Between Rio and Johannesburg the world's nations met in several major conferences under the guidance of the United Nations, including the Monterrey Conference on Finance for Development, as well as the Doha Ministerial Conference. These conferences defined for the world a comprehensive vision for the future of humanity.

10. At the Johannesburg Summit we achieved much in bringing together a rich tapestry of peoples and views in a constructive

search for a common path, towards a world that respects and implements the vision of sustainable development. Johannesburg also confirmed that significant progress has been made towards achieving a global consensus and partnership amongst all the people of our planet.

The Challenges we Face

11. We recognize that poverty eradication, changing consumption and production patterns, and protecting and managing the natural resource base for economic and social development are overarching objectives of, and essential requirements for sustainable development.

12. The deep fault line that divides human society between the rich and the poor and the ever-increasing gap between the developed and developing worlds pose a major threat to global prosperity, security and stability.

13. The global environment continues to suffer. Loss of biodiversity continues, fish stocks continue to be depleted, desertification claims more and more fertile land, the adverse effects of climate change are already evident, natural disasters are more frequent and more devastating and developing countries more vulnerable, and air, water and marine pollution continue to rob millions of a decent life.

14. Globalization has added a new dimension to these challenges. The rapid integration of markets, mobility of capital and significant increases in investment flows around the world have opened new challenges and opportunities for the pursuit of sustainable development. But the benefits and costs of globalization are unevenly distributed, with developing countries facing special difficulties in meeting this challenge.

15. We risk the entrenchment of these global disparities and unless we act in a manner that fundamentally changes their lives, the poor of the world may lose confidence in their representatives and the democratic systems to which we remain committed, seeing their representatives as nothing more than sounding brass or tinkling cymbals.

Our Commitment to Sustainable Development

16. We are determined to ensure that our rich diversity, which is our collective strength, will be used for constructive partnership for change and for the achievement of the common goal of sustainable development.

17. Recognizing the importance of building human solidarity, we urge the promotion of dialogue and cooperation among the world's civilizations and peoples, irrespective of race, disabilities, religion, language, culture and tradition.

18. We welcome the Johannesburg Summit focus on the indivisibility of human dignity and are resolved through decisions on targets, timetables and partnerships to speedily increase access to basic requirements such as clean water, sanitation, adequate shelter, energy, health care, food security and the protection of bio-diversity. At the same time, we will work together to assist one another to have access to financial resources, benefit from the opening of markets, ensure capacity building, use modern technology to bring about development, and make sure that there is technology transfer, human resource development, education and training to banish forever underdevelopment.

19. We reaffirm our pledge to place particular focus on, and give priority attention to, the fight against the worldwide conditions that pose severe threats to the sustainable development of our people. Among these conditions are: chronic hunger; malnutrition;

foreign occupation; armed conflicts; illicit drug problems; organized crime; corruption; natural disasters; illicit arms trafficking; trafficking in persons; terrorism; intolerance and incitement to racial, ethnic, religious and other hatreds; xenophobia; and endemic, communicable and chronic diseases, in particular HIV/AIDS, malaria and tuberculosis.

20. We are committed to ensure that women's empowerment and emancipation, and gender equality are integrated in all activities encompassed within Agenda 21, the Millennium Development Goals and the Johannesburg Plan of Implementation.

21. We recognize the reality that global society has the means and is endowed with the resources to address the challenges of poverty eradication and sustainable development confronting all humanity. Together we will take extra steps to ensure that these available resources are used to the benefit of humanity.

22. In this regard, to contribute to the achievement of our development goals and targets, we urge developed countries that have not done so to make concrete efforts towards the internationally agreed levels of Official Development Assistance.

23. We welcome and support the emergence of stronger regional groupings and alliances, such as the New Partnership for Africa's Development(NEPAD), to promote regional cooperation, improved international co-operation and promote sustainable development.

24. We shall continue to pay special attention to the developmental needs of Small Island Developing States and the Least Developed Countries.

25. We reaffirm the vital role of the indigenous peoples in sustainable development.

26. We recognize sustainable development requires a long-term perspective and broad-based participation in policy formulation,

decision-making and implementation at all levels. As social partners we will continue to work for stable partnerships with all major groups respecting the independent, important roles of each of these.

27. We agree that in pursuit of their legitimate activities the private sector, both large and small companies, have a duty to contribute to the evolution of equitable and sustainable communities and societies.

28. We also agree to provide assistance to increase income generating employment opportunities, taking into account the International Labour Organization(ILO) Declaration of Fundamental Principles and Rights at Work.

29. We agree that there is a need for private sector corporations to enforce corporate accountability. This should take place within a transparent and stable regulatory environment.

30. We undertake to strengthen and improve governance at all levels, for the effective implementation of Agenda 21, the Millennium Development Goals and the Johannesburg Plan of Implementation.

Multilateralism is the Future

31. To achieve our goals of sustainable development, we need more effective, democratic and accountable international and multilateral institutions.

32. We reaffirm our commitment to the principles and purposes of the UN Charter and international law as well as the strengthening of multi-lateralism. We support the leadership role of the United Nations as the most universal and representative organization in the world, which is best placed to promote sustainable development.

33. We further commit ourselves to monitor progress at regular intervals towards the achievement of our sustainable development goals and objectives.

Making it Happen!

34. We are in agreement that this must be an inclusive process, involving all the major groups and governments that participated in the historic Johannesburg Summit.

35. We commit ourselves to act together, united by a common determination to save our planet, promote human development and achieve universal prosperity and peace.

36. We commit ourselves to the Johannesburg Plan of Implementation and to expedite the achievement of the time-bound, socio-economic and environmental targets contained therein.

37. From the African continent, the Cradle of Humankind, we solemnly pledge to the peoples of the world, and the generations that will surely inherit this earth, that we are determined to ensure that our collective hope for sustainable development is realized.

We express our deepest gratitude to the people and the Government of South Africa for their generous hospitality and excellent arrangements made for the World Summit on Sustainable Development.

11. Bali Action Plan (2007)

(FCCC/CP/2007/6/Add.1)
Date: 3-15 December 2007
Link: www.unfccc.int

The Conference of the Parties,

Resolving to urgently enhance implementation of the Convention in order to achieve its ultimate objective in full accordance with its principles and commitments,

Reaffirming that economic and social development and poverty eradication are global priorities,

Responding to the findings of the Fourth Assessment Report of the Intergovernmental Panel on Climate Change that warming of the climate system is unequivocal, and that delay in reducing emissions significantly constrains opportunities to achieve lower stabilization levels and increases the risk of more severe climate change impacts,

Recognizing that deep cuts in global emissions will be required to achieve the ultimate objective of the Convention and emphasizing the urgency1 to address climate change as indicated in the Fourth Assessment Report of the Intergovernmental Panel on Climate Change,

1. Decides to launch a comprehensive process to enable the full, effective and sustained implementation of the Convention through long-term cooperative action, now, up to and beyond 2012, in order to reach an agreed outcome and adopt a decision at its fifteenth session, by addressing, *inter alia*:

(a) A shared vision for long-term cooperative action, including a long-term global goal for emission reductions, to achieve the ultimate objective of the Convention, in accordance with the provisions and principles of the Convention, in particular the principle of common but differentiated responsibilities and respective capabilities, and taking into account social and economic conditions and other relevant factors;

(b) Enhanced national/international action on mitigation of climate change, including, inter alia, consideration of:

(i) Measurable, reportable and verifiable nationally appropriate mitigation commitments or actions, including quantified emission limitation and reduction objectives, by all developed country Parties, while ensuring the comparability of efforts among them, taking into account differences in their national circumstances;

(ii) Nationally appropriate mitigation actions by developing country Parties in the context of sustainable development, supported and enabled by technology, financing and capacity-building, in a measurable, reportable and verifiable manner;

(iii) Policy approaches and positive incentives on issues relating to reducing emissions from deforestation and forest degradation in developing countries; and the role of conservation, sustainable management of forests and enhancement of forest carbon stocks in developing countries;

(iv) Cooperative sectoral approaches and sector-specific actions, in order to enhance implementation of Article 4, paragraph 1(c), of the Convention;

(v) Various approaches, including opportunities for using markets, to enhance the cost-effectiveness of, and to promote, mitigation actions, bearing in mind different circumstances of developed and developing countries;

(vi) Economic and social consequences of response measures;

(vii) Ways to strengthen the catalytic role of the Convention in encouraging multilateral bodies, the public and private sectors and civil society, building on synergies among activities and processes, as a means to support mitigation in a coherent and integrated manner;

(c) Enhanced action on adaptation, including, inter alia, consideration of:

(i) International cooperation to support urgent implementation of adaptation actions, including through vulnerability assessments, prioritization of actions, financial needs assessments, capacity-building and response strategies, integration of adaptation actions into sectoral and national planning, specific projects and programmes, means to incentivize the implementation of adaptation actions, and other ways to enable climate-resilient development and reduce vulnerability of all Parties, taking into account the urgent and immediate needs of developing countries that are particularly vulnerable to the adverse effects of climate change, especially the least developed countries and small island developing States, and further taking into account the needs of countries in Africa affected by drought, desertification and floods;

(ii) Risk management and risk reduction strategies, including risk sharing and transfer mechanisms such as insurance;

(iii) Disaster reduction strategies and means to address loss and damage associated with climate change impacts in developing countries that are particularly vulnerable to the adverse effects of climate change;

(iv) Economic diversification to build resilience;

(v) Ways to strengthen the catalytic role of the Convention in encouraging multilateral bodies, the public and private sectors and civil society, building on synergies among activities and processes, as a means to support adaptation in a coherent and integrated manner;

(d) Enhanced action on technology development and transfer to support action on mitigation and adaptation, including, inter alia, consideration of:

(i) Effective mechanisms and enhanced means for the removal of obstacles to, and provision of financial and other incentives for, scaling up of the development and transfer of technology to developing country Parties in order to promote access to affordable environmentally sound technologies;

(ii) Ways to accelerate deployment, diffusion and transfer of affordable environmentally sound technologies;

(iii) Cooperation on research and development of current, new and innovative technology, including win-win solutions;

(iv) The effectiveness of mechanisms and tools for technology cooperation in specific sectors;

(e) Enhanced action on the provision of financial resources and investment to support action on mitigation and adaptation and technology cooperation, including, inter alia, consideration of:

(i) Improved access to adequate, predictable and sustainable financial resources and financial and technical support, and the provision of new and additional resources, including official and concessional funding for developing country Parties;

(ii) Positive incentives for developing country Parties for the enhanced implementation of national mitigation strategies and adaptation action;

(iii) Innovative means of funding to assist developing country Parties that are particularly vulnerable to the adverse impacts of climate change in meeting the cost of adaptation;

(iv) Means to incentivize the implementation of adaptation actions on the basis of sustainable development policies;

(v) Mobilization of public- and private-sector funding and

investment, including facilitation of carbon-friendly investment choices;

(vi) Financial and technical support for capacity-building in the assessment of the costs of adaptation in developing countries, in particular the most vulnerable ones, to aid in determining their financial needs;

2. Decides that the process shall be conducted under a subsidiary body under the Convention, hereby established and known as the Ad Hoc Working Group on Long-term Cooperative Action under the Convention, that shall complete its work in 2009 and present the outcome of its work to the Conference of the Parties for adoption at its fifteenth session;

3. Agrees that the process shall begin without delay, that the sessions of the group will be scheduled as often as is feasible and necessary to complete the work of the group, where possible in conjunction with sessions of other bodies established under the Convention, and that its sessions may be complemented by workshops and other activities, as required;

4. Decides that the first session of the group shall be held as soon as is feasible and not later than April 2008;

5. Decides that the Chair and Vice-Chair of the group, with one being from a Party included in Annex I to the Convention (Annex I Party) and the other being from a Party not included in Annex I to the Convention (non-Annex I Party), shall alternate annually between an Annex I Party and a non-Annex I Party;

6. Takes note of the proposed schedule of meetings contained in the annex;

7. Instructs the group to develop its work programme at its first session in a coherent and integrated manner;

8. Invites Parties to submit to the secretariat, by 22 February

2008, their views regarding the work programme, taking into account the elements referred to in paragraph 1 above, to be compiled by the secretariat for consideration by the group at its first meeting;

9. Requests the group to report to the Conference of the Parties at its fourteenth session on progress made;

10. Agrees to take stock of the progress made, at its fourteenth session, on the basis of the report by the group;

11. Agrees that the process shall be informed by, inter alia, the best available scientific information, experience in implementation of the Convention and its Kyoto Protocol, and processes thereunder, outputs from other relevant intergovernmental processes and insights from the business and research communities and civil society;

12. Notes that the organization of work of the group will require a significant amount of additional resources to provide for the participation of delegates from Parties eligible to be funded and to provide conference services and substantive support;

13. Strongly urges Parties in a position to do so, in order to facilitate the work of the group, to provide contributions to the Trust Fund for Participation in the UNFCCC Process and the Trust Fund for Supplementary Activities for the purposes referred to in paragraph 12 above and to provide other forms of in kind support such as hosting a session of the group.

Annex [생략]

12. Copenhagen Accord (2009)

(FCCC/CP/2009/L.7) COP 15
Date: 18 December 2009
Link: www.unfccc.int

The Heads of State, Heads of Government, Ministers, and other heads of delegation present at the United Nations Climate Change Conference 2009 in Copenhagen,

In pursuit of the ultimate objective of the Convention as stated in its Article 2,

Being guided by the principles and provisions of the Convention,

Noting the results of work done by the two Ad hoc Working Groups,

Endorsing decision x/CP.15 on the Ad hoc Working Group on Long-term Cooperative Action and decision x/CMP.5 that requests the Ad hoc Working Group on Further Commitments of Annex I Parties under the Kyoto Protocol to continue its work,

Have agreed on this Copenhagen Accord which is operational immediately.

1. We underline that climate change is one of the greatest challenges of our time. We emphasise our strong political will to urgently combat climate change in accordance with the principle of common but differentiated responsibilities and respective capabilities. To achieve the ultimate objective of the Convention to stabilize greenhouse gas concentration in the atmosphere at a level that would prevent dangerous anthropogenic interference with

the climate system, we shall, recognizing the scientific view that the increase in global temperature should be below 2 degrees Celsius, on the basis of equity and in the context of sustainable development, enhance our long-term cooperative action to combat climate change. We recognize the critical impacts of climate change and the potential impacts of response measures on countries particularly vulnerable to its adverse effects and stress the need to establish a comprehensive adaptation programme including international support.

2. We agree that deep cuts in global emissions are required according to science, and as documented by the IPCC Fourth Assessment Report with a view to reduce global emissions so as to hold the increase in global temperature below 2 degrees Celsius, and take action to meet this objective consistent with science and on the basis of equity. We should cooperate in achieving the peaking of global and national emissions as soon as possible, recognizing that the time frame for peaking will be longer in developing countries and bearing in mind that social and economic development and poverty eradication are the first and overriding priorities of developing countries and that a low-emission development strategy is indispensable to sustainable development.

3. Adaptation to the adverse effects of climate change and the potential impacts of response measures is a challenge faced by all countries. Enhanced action and international cooperation on adaptation is urgently required to ensure the implementation of the Convention by enabling and supporting the implementation of adaptation actions aimed at reducing vulnerability and building resilience in developing countries, especially in those that are particularly vulnerable, especially least developed countries, small island developing States and Africa. We agree that developed countries shall provide adequate, predictable and sustainable financial resources, technology and capacity-building to support the

implementation of adaptation action in developing countries.

4. Annex I Parties commit to implement individually or jointly the quantified economywide emissions targets for 2020, to be submitted in the format given in Appendix I by Annex I Parties to the secretariat by 31 January 2010 for compilation in an INF document. Annex I Parties that are Party to the Kyoto Protocol will thereby further strengthen the emissions reductions initiated by the Kyoto Protocol. Delivery of reductions and financing by developed countries will be measured, reported and verified in accordance with existing and any further guidelines adopted by the Conference of the Parties, and will ensure that accounting of such targets and finance is rigorous, robust and transparent.

5. Non-Annex I Parties to the Convention will implement mitigation actions, including those to be submitted to the secretariat by non-Annex I Parties in the format given in Appendix II by 31 January 2010, for compilation in an INF document, consistent with Article 4.1 and Article 4.7 and in the context of sustainable development. Least developed countries and small island developing States may undertake actions voluntarily and on the basis of support. Mitigation actions subsequently taken and envisaged by Non-Annex I Parties, including national inventory reports, shall be communicated through national communications consistent with Article 12.1(b) every two years on the basis of guidelines to be adopted by the Conference of the Parties. Those mitigation actions in national communications or otherwise communicated to the Secretariat will be added to the list in appendix II. Mitigation actions taken by Non-Annex I Parties will be subject to their domestic measurement, reporting and verification the result of which will be reported through their national communications every two years. Non-Annex I Parties will communicate information on the implementation of their actions through National Communications, with provisions for international consultations and analysis under clearly defined

guidelines that will ensure that national sovereignty is respected. Nationally appropriate mitigation actions seeking international support will be recorded in a registry along with relevant technology, finance and capacity building support. Those actions supported will be added to the list in appendix II. These supported nationally appropriate mitigation actions will be subject to international measurement, reporting and verification in accordance with guidelines adopted by the Conference of the Parties.

6. We recognize the crucial role of reducing emission from deforestation and forest degradation and the need to enhance removals of greenhouse gas emission by forests and agree on the need to provide positive incentives to such actions through the immediate establishment of a mechanism including REDD-plus, to enable the mobilization of financial resources from developed countries.

7. We decide to pursue various approaches, including opportunities to use markets, to enhance the cost-effectiveness of, and to promote mitigation actions. Developing countries, especially those with low emitting economies should be provided incentives to continue to develop on a low emission pathway.

8. Scaled up, new and additional, predictable and adequate funding as well as improved access shall be provided to developing countries, in accordance with the relevant provisions of the Convention, to enable and support enhanced action on mitigation, including substantial finance to reduce emissions from deforestation and forest degradation(REDD-plus), adaptation, technology development and transfer and capacity-building, for enhanced implementation of the Convention. The collective commitment by developed countries is to provide new and additional resources, including forestry and investments through international institutions, approaching USD 30 billion for the

period 2010-2012 with balanced allocation between adaptation and mitigation. Funding for adaptation will be prioritized for the most vulnerable developing countries, such as the least developed countries, small island developing States and Africa. In the context of meaningful mitigation actions and transparency on implementation, developed countries commit to a goal of mobilizing jointly USD 100 billion dollars a year by 2020 to address the needs of developing countries. This funding will come from a wide variety of sources, public and private, bilateral and multilateral, including alternative sources of finance. New multilateral funding for adaptation will be delivered through effective and efficient fund arrangements, with a governance structure providing for equal representation of developed and developing countries. A significant portion of such funding should flow through the Copenhagen Green Climate Fund.

9. To this end, a High Level Panel will be established under the guidance of and accountable to the Conference of the Parties to study the contribution of the potential sources of revenue, including alternative sources of finance, towards meeting this goal.

10. We decide that the Copenhagen Green Climate Fund shall be established as an operating entity of the financial mechanism of the Convention to support projects, programme, policies and other activities in developing countries related to mitigation including REDD-plus, adaptation, capacitybuilding, technology development and transfer.

11. In order to enhance action on development and transfer of technology we decide to establish a Technology Mechanism to accelerate technology development and transfer in support of action on adaptation and mitigation that will be guided by a country-driven approach and be based on national circumstances and priorities.

12. We call for an assessment of the implementation of this

Accord to be completed by 2015, including in light of the Convention.s ultimate objective. This would include consideration of strengthening the long-term goal referencing various matters presented by the science, including in relation to temperature rises of 1.5 degrees Celsius.

Appendixes [생략]

Decision 3/CP.15
Amendment to Annex I to the Convention

The Conference of the Parties,

Recalling Articles 15 and 16 of the Convention,

Taking note of the proposal by Malta to amend Annex I to the Convention by adding the name of Malta,1

1. *Decides* to amend Annex I to the Convention by including the name of Malta;

2. *Notes* that in accordance with Article 16, paragraph 4, the entry into force of this amendment to Annex I to the Convention shall be subject to the same procedure as that for the entry into force of annexes to the Convention provided for in Article 16, paragraph 3, of the Convention.

9th plenary meeting
18-19 December 2009

Part Ⅲ. Biodiversity

13. Ramsar Convention on Wetlands of International Importance (1971)

Date: 2 February 1971
In force: 21 December 1975
Contracting Parties: Signatories 18, Parties 160
Korea: 28 July 1997 (No.1415)
Link: www.ramsar.org

The Contracting Parties,

Recognizing the interdependence of Man and his environment;

Considering the fundamental ecological functions of wetlands as regulators of water regimes and as habitats supporting a characteristic flora and fauna, especially waterfowl;

Being Convinced that wetlands constitute a resource of great economic, cultural, scientific, and recreational value, the loss of which would be irreparable;

Desiring to stem the progressive encroachment on and loss of wetlands now and in the future;

Recognizing that waterfowl in their seasonal migrations may transcend frontiers and so should be regarded as an international resource;

Being Confident that the conservation of wetlands and their flora and fauna can be ensured by combining far-sighted national policies with co-ordinated international action;

Have agreed as follows:

Article 1

1. For the purpose of this Convention wetlands are areas of marsh, fen, peatland or water, whether natural or artificial, permanent or temporary, with water that is static or flowing, fresh, brackish or salt, including areas of marine water the depth of which at low tide does not exceed six metres.

2. For the purpose of this Convention waterfowl are birds ecologically dependent on wetlands.

Article 2

1. Each Contracting Party shall designate suitable wetlands within its territory for inclusion in a List of Wetlands of International Importance, hereinafter referred to as "the List" which is maintained by the bureau established under Article 8. The boundaries of each wetland shall be precisely described and also delimited on a map and they may incorporate riparian and coastal zones adjacent to the wetlands, and islands or bodies of marine water deeper than six metres at low tide lying within the wetlands, especially where these have importance as waterfowl habitat.

2. Wetlands should be selected for the List on account of their international significance in terms of ecology, botany, zoology, limnology or hydrology. In the first instance wetlands of international importance to waterfowl at any season should be included.

3. The inclusion of a wetland in the List does not prejudice the exclusive sovereign rights of the Contracting Party in whose territory the wetland is situated.

4. Each Contracting Party shall designate at least one wetland to be included in the List when signing this Convention or when

depositing its instrument of ratification or accession, as provided in Article 9.

5. Any Contracting Party shall have the right to add to the List further wetlands situated within its territory, to extend the boundaries of those wetlands already included by it in the List, or, because of its urgent national interests, to delete or restrict the boundaries of wetlands already included by it in the List and shall, at the earliest possible time, inform the organization or government responsible for the continuing bureau duties specified in Article 8 of any such changes.

6. Each Contracting Party shall consider its international responsibilities for the conservation, management and wise use of migratory stocks of waterfowl, both when designating entries for the List and when exercising its right to change entries in the List relating to wetlands within its territory.

Article 3

1. The Contracting Parties shall formulate and implement their planning so as to promote the conservation of the wetlands included in the List, and as far as possible the wise use of wetlands in their territory.

2. Each Contracting Party shall arrange to be informed at the earliest possible time if the ecological character of any wetland in its territory and included in the List has changed, is changing or is likely to change as the result of technological developments, pollution or other human interference. Information on such changes shall be passed without delay to the organization or government responsible for the continuing bureau duties specified in Article 8.

Article 4

1. Each Contracting Party shall promote the conservation of wetlands and waterfowl by establishing nature reserves on wetlands, whether they are included in the List or not, and provide adequately for their wardening.

2. Where a Contracting Party in its urgent national interest, deletes or restricts the boundaries of a wetland included in the List, it should as far as possible compensate for any loss of wetland resources, and in particular it should create additional nature reserves for waterfowl and for the protection, either in the same area or elsewhere, of an adequate portion of the original habitat.

3. The Contracting Parties shall encourage research and the exchange of data and publications regarding wetlands and their flora and fauna.

4. The Contracting Parties shall endeavour through management to increase waterfowl populations on appropriate wetlands.

5. The Contracting Parties shall promote the training of personnel competent in the fields of wetland research, management and wardening.

Article 5

The Contracting Parties shall consult with each other about implementing obligations arising from the Convention especially in the case of a wetland extending over the territories of more than one Contracting Party or where a water system is shared by Contracting Parties. They shall at the same time endeavour to coordinate and support present and future policies and regulations concerning the conservation of wetlands and their flora and fauna.

Article 6

1. There shall be established a Conference of the Contracting Parties to review and promote the implementation of this Convention. The Bureau referred to in Article 8, paragraph 1, shall convene ordinary meetings of the Conference of the Contracting Parties at intervals of not more than three years, unless the Conference decides otherwise, and extraordinary meetings at the written requests of at least one third of the Contracting Parties. Each ordinary meeting of the Conference of the Contracting Parties shall determine the time and venue of the next ordinary meeting.

2. The Conference of the Contracting Parties shall be competent:

a) to discuss the implementation of this Convention;
b) to discuss additions to and changes in the List;
c) to consider information regarding changes in the ecological character of wetlands included in the List provided in accordance with paragraph 2 of Article 3;
d) to make general or specific recommendations to the Contracting Parties regarding the conservation, management and wise use of wetlands and their flora and fauna;
e) to request relevant international bodies to prepare reports and statistics on matters which are essentially international in character affecting wetlands;
f) to adopt other recommendations, or resolutions, to promote the functioning of this Convention.

3. The Contracting Parties shall ensure that those responsible at all levels for wetlands management shall be informed of, and take into consideration, recommendations of such Conferences concerning the conservation, management and wise use of wetlands and their flora and fauna.

4. The Conference of the Contracting Parties shall adopt rules of

procedure for each of its meetings.

5. The Conference of the Contracting Parties shall establish and keep under review the financial regulations of this Convention. At each of its ordinary meetings, it shall adopt the budget for the next financial period by a two-third majority of Contracting Parties present and voting.

6. Each Contracting Party shall contribute to the budget according to a scale of contributions adopted by unanimity of the Contracting Parties present and voting at a meeting of the ordinary Conference of the Contracting Parties.

Article 7

1. The representatives of the Contracting Parties at such Conferences should include persons who are experts on wetlands or waterfowl by reason of knowledge and experience gained in scientific, administrative or other appropriate capacities.

2. Each of the Contracting Parties represented at a Conference shall have one vote, recommendations, resolutions and decisions being adopted by a simple majority of the Contracting Parties present and voting, unless otherwise provided for in this Convention.

Article 8

1. The International Union for Conservation of Nature and Natural Resources shall perform the continuing bureau duties under this Convention until such time as another organization or government is appointed by a majority of two-thirds of all Contracting Parties.

2. The continuing bureau duties shall be, *inter alia*:

(a) to assist in the convening and organizing of Conferences specified in Article 6;
(b) to maintain the List of Wetlands of International Importance and to be informed by the Contracting Parties of any additions, extensions, deletions or restrictions concerning wetlands included in the List provided in accordance with paragraph 5 of Article 2;
(c) to be informed by the Contracting Parties of any changes in the ecological character of wetlands included in the List provided in accordance with paragraph 2 of Article 3;
(d) to forward notification of any alterations to the List, or changes in character of wetlands included therein, to all Contracting Parties and to arrange for these matters to be discussed at the next Conference;
(e) to make known to the Contracting Party concerned, the recommendations of the Conferences in respect of such alterations to the List or of changes in the character of wetlands included therein.

Article 9

1. This Convention shall remain open for signature indefinitely.

2. Any member of the United Nations or of one of the Specialized Agencies or of the International Atomic Energy Agency or Party to the Statute of the International Court of Justice may become a Party to this Convention by:

(a) signature without reservation as to ratification;
(b) signature subject to ratification followed by ratification;
(c) accession.

3. Ratification or accession shall be effected by the deposit of an instrument of ratification or accession with the Director-General of the United Nations Educational, Scientific and Cultural Organ-

ization(hereinafter referred to as "the Depositary").

Article 10

1. This Convention shall enter into force four months after seven States have become Parties to this Convention in accordance with paragraph 2 of Article 9.

2. Thereafter this Convention shall enter into force for each Contracting Party four months after the day of its signature without reservation as to ratification, or its deposit of an instrument of ratification or accession.

Article 10 bis

1. This Convention may be amended at a meeting of the Contracting Parties convened for that purpose in accordance with this article.

2. Proposals for amendment may be made by any Contracting Party.

3. The text of any proposed amendment and the reasons for it shall be communicated to the organization or government performing the continuing bureau duties under the Convention (hereinafter referred to as "the Bureau") and shall promptly be communicated by the Bureau to all Contracting Parties. Any comments on the text by the Contracting Parties shall be communicated to the Bureau within three months of the date on which the amendments were communicated to the Contracting Parties by the Bureau. The Bureau shall, immediately after the last day for submission of comments, communicate to the Contracting Parties all comments submitted by that day.

4. A meeting of Contracting Parties to consider an amendment communicated in accordance with paragraph 3 shall be convened

by the Bureau upon the written request of one third of the Contracting Parties. The Bureau shall consult the Parties concerning the time and venue of the meeting.

5. Amendments shall be adopted by a two-thirds majority of the Contracting Parties present and voting.

6. An amendment adopted shall enter into force for the Contracting Parties which have accepted it on the first day of the fourth month following the date on which two thirds of the Contracting Parties have deposited an instrument of acceptance with the Depositary. For each Contracting Party which deposits an instrument of acceptance after the date on which two thirds of the Contracting Parties have deposited an instrument of acceptance, the amendment shall enter into force on the first day of the fourth month following the date of the deposit of its instrument of acceptance.

Article 11

1. This Convention shall continue in force for an indefinite period.

2. Any Contracting Party may denounce this Convention after a period of five years from the date on which it entered into force for that party by giving written notice thereof to the Depositary. Denunciation shall take effect four months after the day on which notice thereof is received by the Depositary.

Article 12

1. The Depositary shall inform all States that have signed and acceded to this Convention as soon as possible of:

(a) signatures to the Convention;
(b) deposits of instruments of ratification of this Convention;
(c) deposits of instruments of accession to this Convention;

(d) the date of entry into force of this Convention;

(e) notifications of denunciation of this Convention.

2. When this Convention has entered into force, the Depositary shall have it registered with the Secretariat of the United Nations in accordance with Article 102 of the Charter.

IN WITNESS WHEREOF, the undersigned, being duly authorized to that effect, have signed this Convention.

DONE at Ramsar this 2nd day of February 1971, in a single original in the English, French, German and Russian languages, all texts being equally authentic[1)] which shall be deposited with the Depositary which shall send true copies thereof to all Contracting Parties.

---

* Pursuant to the Final Act of the Conference to conclude the Protocol, the Depositary provided the second Conference of the Contracting Parties with official versions of the Convention in the Arabic, Chinese and Spanish languages, prepared in consultation with interested Governments and with the assistance of the Bureau.

14. Convention on International Trade in Endangered Species of Wild Fauna and Flora (1973)

Date: 3 March 1973
In force: 1 July 1975
Contracting Parties: 175
Korea: 7 October 1993 (No.1194)
Link: www.cites.org

The Contracting States,

Recognizing that wild fauna and flora in their many beautiful and varied forms are an irreplaceable part of the natural systems of the earth which must be protected for this and the generations to come;

Conscious of the ever-growing value of wild fauna and flora from aesthetic, scientific, cultural, recreational and economic points of view;

Recognizing that peoples and States are and should be the best protectors of their own wild fauna and flora;

Recognizing, in addition, that international co-operation is essential for the protection of certain species of wild fauna and flora against over-exploitation through international trade;

Convinced of the urgency of taking appropriate measures to this end;

Have agreed as follows:

Article I
Definitions

For the purpose of the present Convention, unless the context otherwise requires:

(a) "Species" means any species, subspecies, or geographically separate population thereof;

(b) "Specimen" means:

(i) any animal or plant, whether alive or dead;

(ii) in the case of an animal: for species included in Appendices I and II, any readily recognizable part or derivative thereof; and for species included in Appendix III, any readily recognizable part or derivative thereof specified in Appendix III in relation to the species; and

(iii) in the case of a plant: for species included in Appendix I, any readily recognizable part or derivative thereof; and for species included in Appendices II and III, any readily recognizable part or derivative thereof specified in Appendices II and III in relation to the species;

(c) "Trade" means export, re-export, import and introduction from the sea;

(d) "Re-export" means export of any specimen that has previously been imported;

(e) "Introduction from the sea" means transportation into a State of specimens of any species which were taken in the marine environment not under the jurisdiction of any State;

(f) "Scientific Authority" means a national scientific authority designated in accordance with Article IX;

(g) "Management Authority" means a national management authority designated in accordance with Article IX;

(h) "Party" means a State for which the present Convention has entered into force.

Article II
Fundamental Principles

1. Appendix I shall include all species threatened with extinction which are or may be affected by trade. Trade in specimens of these species must be subject to particularly strict regulation in order not to endanger further their survival and must only be authorized in exceptional circumstances.

2. Appendix II shall include:

(a) all species which although not necessarily now threatened with extinction may become so unless trade in specimens of such species is subject to strict regulation in order to avoid utilization incompatible with their survival; and
(b) other species which must be subject to regulation in order that trade in specimens of certain species referred to in sub-paragraph (a) of this paragraph may be brought under effective control.

3. Appendix III shall include all species which any Party identifies as being subject to regulation within its jurisdiction for the purpose of preventing or restricting exploitation, and as needing the co-operation of other Parties in the control of trade.

4. The Parties shall not allow trade in specimens of species included in Appendices I, II and III except in accordance with the provisions of the present Convention.

Article III
Regulation of Trade in Specimens of Species Included in Appendix I

1. All trade in specimens of species included in Appendix I shall be in accordance with the provisions of this Article.

2. The export of any specimen of a species included in Appendix I

shall require the prior grant and presentation of an export permit. An export permit shall only be granted when the following conditions have been met:

(a) a Scientific Authority of the State of export has advised that such export will not be detrimental to the survival of that species;
(b) a Management Authority of the State of export is satisfied that the specimen was not obtained in contravention of the laws of that State for the protection of fauna and flora;
(c) a Management Authority of the State of export is satisfied that any living specimen will be so prepared and shipped as to minimize the risk of injury, damage to health or cruel treatment; and
(d) a Management Authority of the State of export is satisfied that an import permit has been granted for the specimen.

3. The import of any specimen of a species included in Appendix I shall require the prior grant and presentation of an import permit and either an export permit or a re-export certificate. An import permit shall only be granted when the following conditions have been met:

(a) a Scientific Authority of the State of import has advised that the import will be for purposes which are not detrimental to the survival of the species involved;
(b) a Scientific Authority of the State of import is satisfied that the proposed recipient of a living specimen is suitably equipped to house and care for it; and
(c) a Management Authority of the State of import is satisfied that the specimen is not to be used for primarily commercial purposes.

4. The re-export of any specimen of a species included in Appendix I shall require the prior grant and presentation of a re-export certificate. A re-export certificate shall only be granted

when the following conditions have been met:

(a) a Management Authority of the State of re-export is satisfied that the specimen was imported into that State in accordance with the provisions of the present Convention;
(b) a Management Authority of the State of re-export is satisfied that any living specimen will be so prepared and shipped as to minimize the risk of injury, damage to health or cruel treatment; and
(c) a Management Authority of the State of re-export is satisfied that an import permit has been granted for any living specimen.

5. The introduction from the sea of any specimen of a species included in Appendix I shall require the prior grant of a certificate from a Management Authority of the State of introduction. A certificate shall only be granted when the following conditions have been met:

(a) a Scientific Authority of the State of introduction advises that the introduction will not be detrimental to the survival of the species involved;
(b) a Management Authority of the State of introduction is satisfied that the proposed recipient of a living specimen is suitably equipped to house and care for it; and
(c) a Management Authority of the State of introduction is satisfied that the specimen is not to be used for primarily commercial purposes.

Article IV
Regulation of Trade in Specimens of Species Included in Appendix II

1. All trade in specimens of species included in Appendix II shall be in accordance with the provisions of this Article.

2. The export of any specimen of a species included in Appendix II

shall require the prior grant and presentation of an export permit. An export permit shall only be granted when the following conditions have been met:

(a) a Scientific Authority of the State of export has advised that such export will not be detrimental to the survival of that species;
(b) a Management Authority of the State of export is satisfied that the specimen was not obtained in contravention of the laws of that State for the protection of fauna and flora; and
(c) a Management Authority of the State of export is satisfied that any living specimen will be so prepared and shipped as to minimize the risk of injury, damage to health or cruel treatment.

3. A Scientific Authority in each Party shall monitor both the export permits granted by that State for specimens of species included in Appendix II and the actual exports of such specimens. Whenever a Scientific Authority determines that the export of specimens of any such species should be limited in order to maintain that species throughout its range at a level consistent with its role in the ecosystems in which it occurs and well above the level at which that species might become eligible for inclusion in Appendix I, the Scientific Authority shall advise the appropriate Management Authority of suitable measures to be taken to limit the grant of export permits for specimens of that species.

4. The import of any specimen of a species included in Appendix II shall require the prior presentation of either an export permit or a re-export certificate.

5. The re-export of any specimen of a species included in Appendix II shall require the prior grant and presentation of a re-export certificate. A re-export certificate shall only be granted when the following conditions have been met:

(a) a Management Authority of the State of re-export is satisfied that the specimen was imported into that State in accordance with the provisions of the present Convention; and

(b) a Management Authority of the State of re-export is satisfied that any living specimen will be so prepared and shipped as to minimize the risk of injury, damage to health or cruel treatment.

6. The introduction from the sea of any specimen of a species included in Appendix II shall require the prior grant of a certificate from a Management Authority of the State of introduction. A certificate shall only be granted when the following conditions have been met:

(a) a Scientific Authority of the State of introduction advises that the introduction will not be detrimental to the survival of the species involved; and

(b) a Management Authority of the State of introduction is satisfied that any living specimen will be so handled as to minimize the risk of injury, damage to health or cruel treatment.

7. Certificates referred to in paragraph 6 of this Article may be granted on the advice of a Scientific Authority, in consultation with other national scientific authorities or, when appropriate, international scientific authorities, in respect of periods not exceeding one year for total numbers of specimens to be introduced in such periods.

Article V

Regulation of Trade in Specimens of Species Included in Appendix III

1. All trade in specimens of species included in Appendix III shall be in accordance with the provisions of this Article.

2. The export of any specimen of a species included in Appendix III from any State which has included that species in Appendix III shall require the prior grant and presentation of an export permit. An export permit shall only be granted when the following conditions have been met:

(a) a Management Authority of the State of export is satisfied that the specimen was not obtained in contravention of the laws of that State for the protection of fauna and flora; and
(b) a Management Authority of the State of export is satisfied that any living specimen will be so prepared and shipped as to minimize the risk of injury, damage to health or cruel treatment.

3. The import of any specimen of a species included in Appendix III shall require, except in circumstances to which paragraph 4 of this Article applies, the prior presentation of a certificate of origin and, where the import is from a State which has included that species in Appendix III, an export permit.

4. In the case of re-export, a certificate granted by the Management Authority of the State of re-export that the specimen was processed in that State or is being re-exported shall be accepted by the State of import as evidence that the provisions of the present Convention have been complied with in respect of the specimen concerned.

Article VI
Permits and Certificates

1. Permits and certificates granted under the provisions of Articles III, IV, and V shall be in accordance with the provisions of this Article.

2. An export permit shall contain the information specified in the model set forth in Appendix IV, and may only be used for

export within a period of six months from the date on which it was granted.

3. Each permit or certificate shall contain the title of the present Convention, the name and any identifying stamp of the Management Authority granting it and a control number assigned by the Management Authority.

4. Any copies of a permit or certificate issued by a Management Authority shall be clearly marked as copies only and no such copy may be used in place of the original, except to the extent endorsed thereon.

5. A separate permit or certificate shall be required for each consignment of specimens.

6. A Management Authority of the State of import of any specimen shall cancel and retain the export permit or re-export certificate and any corresponding import permit presented in respect of the import of that specimen.

7. Where appropriate and feasible a Management Authority may affix a mark upon any specimen to assist in identifying the specimen. For these purposes "mark" means any indelible imprint, lead seal or other suitable means of identifying a specimen, designed in such a way as to render its imitation by unauthorized persons as difficult as possible.

Article VII
Exemptions and Other Special Provisions Relating to Trade

1. The provisions of Articles III, IV and V shall not apply to the transit or transhipment of specimens through or in the territory of a Party while the specimens remain in Customs control.

2. Where a Management Authority of the State of export or re-export is satisfied that a specimen was acquired before the

provisions of the present Convention applied to that specimen, the provisions of Articles III, IV and V shall not apply to that specimen where the Management Authority issues a certificate to that effect.

3. The provisions of Articles III, IV and V shall not apply to specimens that are personal or household effects. This exemption shall not apply where:

(a) in the case of specimens of a species included in Appendix I, they were acquired by the owner outside his State of usual residence, and are being imported into that State; or
(b) in the case of specimens of species included in Appendix II:
  (i) they were acquired by the owner outside his State of usual residence and in a State where removal from the wild occurred;
  (ii) they are being imported into the owner's State of usual residence; and
  (iii) the State where removal from the wild occurred requires the prior grant of export permits before any export of such specimens; unless a Management Authority is satisfied that the specimens were acquired before the provisions of the present Convention applied to such specimens.

4. Specimens of an animal species included in Appendix I bred in captivity for commercial purposes, or of a plant species included in Appendix I artificially propagated for commercial purposes, shall be deemed to be specimens of species included in Appendix II.

5. Where a Management Authority of the State of export is satisfied that any specimen of an animal species was bred in captivity or any specimen of a plant species was artificially propagated, or is a part of such an animal or plant or was derived therefrom, a certificate by that Management Authority to that effect shall be accepted in lieu of any of the permits or

certificates required under the provisions of Article III, IV or V.

6. The provisions of Articles III, IV and V shall not apply to the non-commercial loan, donation or exchange between scientists or scientific institutions registered by a Management Authority of their State, of herbarium specimens, other preserved, dried or embedded museum specimens, and live plant material which carry a label issued or approved by a Management Authority.

7. A Management Authority of any State may waive the requirements of Articles III, IV and V and allow the movement without permits or certificates of specimens which form part of a travelling zoo, circus, menagerie, plant exhibition or other travelling exhibition provided that:

(a) the exporter or importer registers full details of such specimens with that Management Authority;
(b) the specimens are in either of the categories specified in paragraph 2 or 5 of this Article; and
(c) the Management Authority is satisfied that any living specimen will be so transported and cared for as to minimize the risk of injury, damage to health or cruel treatment.

Article VIII
Measures to Be Taken by the Parties

1. The Parties shall take appropriate measures to enforce the provisions of the present Convention and to prohibit trade in specimens in violation thereof. These shall include measures:

(a) to penalize trade in, or possession of, such specimens, or both; and
(b) to provide for the confiscation or return to the State of export of such specimens.

2. In addition to the measures taken under paragraph 1 of this Article, a Party may, when it deems it necessary, provide for any method of internal reimbursement for expenses incurred as a result of the confiscation of a specimen traded in violation of the measures taken in the application of the provisions of the present Convention.

3. As far as possible, the Parties shall ensure that specimens shall pass through any formalities required for trade with a minimum of delay. To facilitate such passage, a Party may designate ports of exit and ports of entry at which specimens must be presented for clearance. The Parties shall ensure further that all living specimens, during any period of transit, holding or shipment, are properly cared for so as to minimize the risk of injury, damage to health or cruel treatment.

4. Where a living specimen is confiscated as a result of measures referred to in paragraph 1 of this Article:

(a) the specimen shall be entrusted to a Management Authority of the State of confiscation;
(b) the Management Authority shall, after consultation with the State of export, return the specimen to that State at the expense of that State, or to a rescue centre or such other place as the Management Authority deems appropriate and consistent with the purposes of the present Convention; and
(c) the Management Authority may obtain the advice of a Scientific Authority, or may, whenever it considers it desirable, consult the Secretariat in order to facilitate the decision under sub-paragraph (b) of this paragraph, including the choice of a rescue centre or other place.

5. A rescue centre as referred to in paragraph 4 of this Article means an institution designated by a Management Authority to look after the welfare of living specimens, particularly those that have been confiscated.

6. Each Party shall maintain records of trade in specimens of species included in Appendices I, II and III which shall cover:

(a) the names and addresses of exporters and importers; and
(b) the number and type of permits and certificates granted; the States with which such trade occurred; the numbers or quantities and types of specimens, names of species as included in Appendices I, II and III and, where applicable, the size and sex of the specimens in question.

7. Each Party shall prepare periodic reports on its implementation of the present Convention and shall transmit to the Secretariat:

(a) an annual report containing a summary of the information specified in sub-paragraph (b) of paragraph 6 of this Article; and
(b) a biennial report on legislative, regulatory and administrative measures taken to enforce the provisions of the present Convention.

8. The information referred to in paragraph 7 of this Article shall be available to the public where this is not inconsistent with the law of the Party concerned.

Article IX
Management and Scientific Authorities

1. Each Party shall designate for the purposes of the present Convention:

(a) one or more Management Authorities competent to grant permits or certificates on behalf of that Party; and
(b) one or more Scientific Authorities.

2. A State depositing an instrument of ratification, acceptance, approval or accession shall at that time inform the Depositary Government of the name and address of the Management

Authority authorized to communicate with other Parties and with the Secretariat.

3. Any changes in the designations or authorizations under the provisions of this Article shall be communicated by the Party concerned to the Secretariat for transmission to all other Parties.

4. Any Management Authority referred to in paragraph 2 of this Article shall, if so requested by the Secretariat or the Management Authority of another Party, communicate to it impression of stamps, seals or other devices used to authenticate permits or certificates.

Article X
Trade with States not Party to the Convention

Where export or re-export is to, or import is from, a State not a Party to the present Convention, comparable documentation issued by the competent authorities in that State which substantially conforms with the requirements of the present Convention for permits and certificates may be accepted in lieu thereof by any Party.

Article XI
Conference of the Parties

1. The Secretariat shall call a meeting of the Conference of the Parties not later than two years after the entry into force of the present Convention.

2. Thereafter the Secretariat shall convene regular meetings at least once every two years, unless the Conference decides otherwise, and extraordinary meetings at any time on the written request of at least one-third of the Parties.

3. At meetings, whether regular or extraordinary, the Parties shall

review the implementation of the present Convention and may:

(a) make such provision as may be necessary to enable the Secretariat to carry out its duties, and adopt financial provisions;
(b) consider and adopt amendments to Appendices I and II in accordance with Article XV;
(c) review the progress made towards the restoration and conservation of the species included in Appendices I, II and III;
(d) receive and consider any reports presented by the Secretariat or by any Party; and
(e) where appropriate, make recommendations for improving the effectiveness of the present Convention.

4. At each regular meeting, the Parties may determine the time and venue of the next regular meeting to be held in accordance with the provisions of paragraph 2 of this Article.

5. At any meeting, the Parties may determine and adopt rules of procedure for the meeting.

6. The United Nations, its Specialized Agencies and the International Atomic Energy Agency, as well as any State not a Party to the present Convention, may be represented at meetings of the Conference by observers, who shall have the right to participate but not to vote.

7. Any body or agency technically qualified in protection, conservation or management of wild fauna and flora, in the following categories, which has informed the Secretariat of its desire to be represented at meetings of the Conference by observers, shall be admitted unless at least one-third of the Parties present object:

(a) international agencies or bodies, either governmental or non-governmental, and national governmental agencies and bodies; and

(b) national non-governmental agencies or bodies which have been approved for this purpose by the State in which they are located. Once admitted, these observers shall have the right to participate but not to vote.

Article XII
The Secretariat

1. Upon entry into force of the present Convention, a Secretariat shall be provided by the Executive Director of the United Nations Environment Programme. To the extent and in the manner he considers appropriate, he may be assisted by suitable inter-governmental or non-governmental international or national agencies and bodies technically qualified in protection, conservation and management of wild fauna and flora.

2. The functions of the Secretariat shall be:

(a) to arrange for and service meetings of the Parties;

(b) to perform the functions entrusted to it under the provisions of Articles XV and XVI of the present Convention;

(c) to undertake scientific and technical studies in accordance with programmes authorized by the Conference of the Parties as will contribute to the implementation of the present Convention, including studies concerning standards for appropriate preparation and shipment of living specimens and the means of identifying specimens;

(d) to study the reports of Parties and to request from Parties such further information with respect thereto as it deems necessary to ensure implementation of the present Convention;

(e) to invite the attention of the Parties to any matter pertaining to the aims of the present Convention;

(f) to publish periodically and distribute to the Parties current editions of Appendices I, II and III together with any information which will facilitate identification of specimens of species included in those Appendices;

(g) to prepare annual reports to the Parties on its work and on the implementation of the present Convention and such other reports as meetings of the Parties may request;

(h) to make recommendations for the implementation of the aims and provisions of the present Convention, including the exchange of information of a scientific or technical nature;

(i) to perform any other function as may be entrusted to it by the Parties.

Article XIII
International Measures

1. When the Secretariat in the light of information received is satisfied that any species included in Appendix I or II is being affected adversely by trade in specimens of that species or that the provisions of the present Convention are not being effectively implemented, it shall communicate such information to the authorized Management Authority of the Party or Parties concerned.

2. When any Party receives a communication as indicated in paragraph 1 of this Article, it shall, as soon as possible, inform the Secretariat of any relevant facts insofar as its laws permit and, where appropriate, propose remedial action. Where the Party considers that an inquiry is desirable, such inquiry may be carried out by one or more persons expressly authorized by the Party.

3. The information provided by the Party or resulting from any inquiry as specified in paragraph 2 of this Article shall be reviewed by the next Conference of the Parties which may make whatever recommendations it deems appropriate.

Article XIV
Effect on Domestic Legislation and International Conventions

1. The provisions of the present Convention shall in no way affect the right of Parties to adopt:

(a) stricter domestic measures regarding the conditions for trade, taking, possession or transport of specimens of species included in Appendices I, II and III, or the complete prohibition thereof; or
(b) domestic measures restricting or prohibiting trade, taking, possession or transport of species not included in Appendix I, II or III.

2. The provisions of the present Convention shall in no way affect the provisions of any domestic measures or the obligations of Parties deriving from any treaty, convention, or international agreement relating to other aspects of trade, taking, possession or transport of specimens which is in force or subsequently may enter into force for any Party including any measure pertaining to the Customs, public health, veterinary or plant quarantine fields.

3. The provisions of the present Convention shall in no way affect the provisions of, or the obligations deriving from, any treaty, convention or international agreement concluded or which may be concluded between States creating a union or regional trade agreement establishing or maintaining a common external Customs control and removing Customs control between the parties thereto insofar as they relate to trade among the States members of that union or agreement.

4. A State party to the present Convention, which is also a party to any other treaty, convention or international agreement which is in force at the time of the coming into force of the present Convention and under the provisions of which protection is afforded to marine species included in Appendix II, shall be

relieved of the obligations imposed on it under the provisions of the present Convention with respect to trade in specimens of species included in Appendix II that are taken by ships registered in that State and in accordance with the provisions of such other treaty, convention or international agreement.

5. Notwithstanding the provisions of Articles III, IV and V, any export of a specimen taken in accordance with paragraph 4 of this Article shall only require a certificate from a Management Authority of the State of introduction to the effect that the specimen was taken in accordance with the provisions of the other treaty, convention or international agreement in question.

6. Nothing in the present Convention shall prejudice the codification and development of the law of the sea by the United Nations Conference on the Law of the Sea convened pursuant to Resolution 2750 C(XXV) of the General Assembly of the United Nations nor the present or future claims and legal views of any State concerning the law of the sea and the nature and extent of coastal and flag State jurisdiction.

Article XV
Amendments to Appendices I and II

1. The following provisions shall apply in relation to amendments to Appendices I and II at meetings of the Conference of the Parties:

(a) Any Party may propose an amendment to Appendix I or II for consideration at the next meeting. The text of the proposed amendment shall be communicated to the Secretariat at least 150 days before the meeting. The Secretariat shall consult the other Parties and interested bodies on the amendment in accordance with the provisions of sub-paragraphs (b) and (c) of paragraph 2 of this Article and shall communicate the response to all Parties not later than 30 days before the meeting.

(b) Amendments shall be adopted by a two-thirds majority of Parties present and voting. For these purposes "Parties present and voting" means Parties present and casting an affirmative or negative vote. Parties abstaining from voting shall not be counted among the two-thirds required for adopting an amendment.

(c) Amendments adopted at a meeting shall enter into force 90 days after that meeting for all Parties except those which make a reservation in accordance with paragraph 3 of this Article.

2. The following provisions shall apply in relation to amendments to Appendices I and II between meetings of the Conference of the Parties:

(a) Any Party may propose an amendment to Appendix I or II for consideration between meetings by the postal procedures set forth in this paragraph.

(b) For marine species, the Secretariat shall, upon receiving the text of the proposed amendment, immediately communicate it to the Parties. It shall also consult inter-governmental bodies having a function in relation to those species especially with a view to obtaining scientific data these bodies may be able to provide and to ensuring co-ordination with any conservation measures enforced by such bodies. The Secretariat shall communicate the views expressed and data provided by these bodies and its own findings and recommendations to the Parties as soon as possible.

(c) For species other than marine species, the Secretariat shall, upon receiving the text of the proposed amendment, immediately communicate it to the Parties, and, as soon as possible thereafter, its own recommendations.

(d) Any Party may, within 60 days of the date on which the

Secretariat communicated its recommendations to the Parties under sub-paragraph (b) or (c) of this paragraph, transmit to the Secretariat any comments on the proposed amendment together with any relevant scientific data and information.

(e) The Secretariat shall communicate the replies received together with its own recommendations to the Parties as soon as possible.

(f) If no objection to the proposed amendment is received by the Secretariat within 30 days of the date the replies and recommendations were communicated under the provisions of sub-paragraph (e) of this paragraph, the amendment shall enter into force 90 days later for all Parties except those which make a reservation in accordance with paragraph 3 of this Article.

(g) If an objection by any Party is received by the Secretariat, the proposed amendment shall be submitted to a postal vote in accordance with the provisions of sub-paragraphs (h) , (i) and (j) of this paragraph.

(h) The Secretariat shall notify the Parties that notification of objection has been received.

(i) Unless the Secretariat receives the votes for, against or in abstention from at least one-half of the Parties within 60 days of the date of notification under sub-paragraph (h) of this paragraph, the proposed amendment shall be referred to the next meeting of the Conference for further consideration.

(j) Provided that votes are received from one-half of the Parties, the amendment shall be adopted by a two-thirds majority of Parties casting an affirmative or negative vote.

(k) The Secretariat shall notify all Parties of the result of the vote.

(l) If the proposed amendment is adopted it shall enter into force 90 days after the date of the notification by the Secretariat of its acceptance for all Parties except those which make a reservation in accordance with paragraph 3 of this Article.

3. During the period of 90 days provided for by sub-paragraph (c) of paragraph 1 or sub-paragraph (l) of paragraph 2 of this Article any Party may by notification in writing to the Depositary Government make a reservation with respect to the amendment. Until such reservation is withdrawn the Party shall be treated as a State not a Party to the present Convention with respect to trade in the species concerned.

Article XVI
Appendix III and Amendments thereto

1. Any Party may at any time submit to the Secretariat a list of species which it identifies as being subject to regulation within its jurisdiction for the purpose mentioned in paragraph 3 of Article II. Appendix III shall include the names of the Parties submitting the species for inclusion therein, the scientific names of the species so submitted, and any parts or derivatives of the animals or plants concerned that are specified in relation to the species for the purposes of sub-paragraph (b) of Article I.

2. Each list submitted under the provisions of paragraph 1 of this Article shall be communicated to the Parties by the Secretariat as soon as possible after receiving it. The list shall take effect as part of Appendix III 90 days after the date of such communication. At any time after the communication of such list, any Party may by notification in writing to the Depositary Government enter a reservation with respect to any species or any parts or derivatives, and until such reservation is withdrawn, the State shall be treated as a State not a Party to the present Convention with respect to trade in the species or part or derivative concerned.

3. A Party which has submitted a species for inclusion in

Appendix III may withdraw it at any time by notification to the Secretariat which shall communicate the withdrawal to all Parties. The withdrawal shall take effect 30 days after the date of such communication.

4. Any Party submitting a list under the provisions of paragraph 1 of this Article shall submit to the Secretariat a copy of all domestic laws and regulations applicable to the protection of such species, together with any interpretations which the Party may deem appropriate or the Secretariat may request. The Party shall, for as long as the species in question is included in Appendix III, submit any amendments of such laws and regulations or any interpretations as they are adopted.

Article XVII
Amendment of the Convention

1. An extraordinary meeting of the Conference of the Parties shall be convened by the Secretariat on the written request of at least one-third of the Parties to consider and adopt amendments to the present Convention. Such amendments shall be adopted by a two-thirds majority of Parties present and voting. For these purposes "Parties present and voting" means Parties present and casting an affirmative or negative vote. Parties abstaining from voting shall not be counted among the two-thirds required for adopting an amendment.

2. The text of any proposed amendment shall be communicated by the Secretariat to all Parties at least 90 days before the meeting.

3. An amendment shall enter into force for the Parties which have accepted it 60 days after two-thirds of the Parties have deposited an instrument of acceptance of the amendment with the Depositary Government. Thereafter, the amendment shall enter

into force for any other Party 60 days after that Party deposits its instrument of acceptance of the amendment.

Article XVIII
Resolution of Disputes

1. Any dispute which may arise between two or more Parties with respect to the interpretation or application of the provisions of the present Convention shall be subject to negotiation between the Parties involved in the dispute.

2. If the dispute can not be resolved in accordance with paragraph 1 of this Article, the Parties may, by mutual consent, submit the dispute to arbitration, in particular that of the Permanent Court of Arbitration at The Hague, and the Parties submitting the dispute shall be bound by the arbitral decision.

Article XIX
Signature

The present Convention shall be open for signature at Washington until 30th April 1973 and thereafter at Berne until 31st December 1974.

Article XX
Ratification, Acceptance, Approval

The present Convention shall be subject to ratification, acceptance or approval. Instruments of ratification, acceptance or approval shall be deposited with the Government of the Swiss Confederation which shall be the Depositary Government.

Article XXI
Accession

The present Convention shall be open indefinitely for accession. Instruments of accession shall be deposited with the Depositary Government.

Article XXII
Entry into Force

1. The present Convention shall enter into force 90 days after the date of deposit of the tenth instrument of ratification, acceptance, approval or accession, with the Depositary Government.

2. For each State which ratifies, accepts or approves the present Convention or accedes thereto after the deposit of the tenth instrument of ratification, acceptance, approval or accession, the present Convention shall enter into force 90 days after the deposit by such State of its instrument of ratification, acceptance, approval or accession.

Article XXIII
Reservations

1. The provisions of the present Convention shall not be subject to general reservations. Specific reservations may be entered in accordance with the provisions of this Article and Articles XV and XVI.

2. Any State may, on depositing its instrument of ratification, acceptance, approval or accession, enter a specific reservation with regard to:

(a) any species included in Appendix I, II or III; or
(b) any parts or derivatives specified in relation to a species included in Appendix III.

3. Until a Party withdraws its reservation entered under the provisions of this Article, it shall be treated as a State not a Party to the present Convention with respect to trade in the particular species or parts or derivatives specified in such reservation.

Article XXIV
Denunciation

Any Party may denounce the present Convention by written notification to the Depositary Government at any time. The denunciation shall take effect twelve months after the Depositary Government has received the notification.

Article XXV
Depositary

1. The original of the present Convention, in the Chinese, English, French, Russian and Spanish languages, each version being equally authentic, shall be deposited with the Depositary Government, which shall transmit certified copies thereof to all States that have signed it or deposited instruments of accession to it.

2. The Depositary Government shall inform all signatory and acceding States and the Secretariat of signatures, deposit of instruments of ratification, acceptance, approval or accession, entry into force of the present Convention, amendments thereto, entry and withdrawal of reservations and notifications of denunciation.

3. As soon as the present Convention enters into force, a certified copy thereof shall be transmitted by the Depositary Government to the Secretariat of the United Nations for registration and publication in accordance with Article 102 of the Charter of the United Nations.

IN WITNESS WHEREOF the undersigned Plenipotentiaries, being duly authorized to that effect, have signed the present Convention.

DONE at Washington this third day of March, One Thousand Nine Hundred and Seventy-three.

15. Convention on Biological Diversity (1992)

Date: 5 June 1992
In force: 29 December 1993
Contracting Parties: Signatories 168, Parties 193
Korea: 1 January 1995 (No. 1264)
Link: www.cbd.int

Preamble

The Contracting Parties,

Conscious of the intrinsic value of biological diversity and of the ecological, genetic, social, economic, scientific, educational, cultural, recreational and aesthetic values of biological diversity and its components,

Conscious also of the importance of biological diversity for evolution and for maintaining life sustaining systems of the biosphere,

Affirming that the conservation of biological diversity is a common concern of humankind,

Reaffirming that States have sovereign rights over their own biological resources,

Reaffirming also that States are responsible for conserving their biological diversity and for using their biological resources in a sustainable manner,

Concerned that biological diversity is being significantly reduced by certain human activities,

Aware of the general lack of information and knowledge regarding biological diversity and of the urgent need to develop scientific, technical and institutional capacities to provide the basic understanding upon which to plan and implement appropriate measures,

Noting that it is vital to anticipate, prevent and attack the causes of significant reduction or loss of biological diversity at source,

Noting also that where there is a threat of significant reduction or loss of biological diversity, lack of full scientific certainty should not be used as a reason for postponing measures to avoid or minimize such a threat,

Noting further that the fundamental requirement for the conservation of biological diversity is the *in-situ* conservation of ecosystems and natural habitats and the maintenance and recovery of viable populations of species in their natural surroundings,

Noting further that *ex-situ* measures, preferably in the country of origin, also have an important role to play,

Recognizing the close and traditional dependence of many indigenous and local communities embodying traditional lifestyles on biological resources, and the desirability of sharing equitably benefits arising from the use of traditional knowledge, innovations and practices relevant to the conservation of biological diversity and the sustainable use of its components,

Recognizing also the vital role that women play in the conservation and sustainable use of biological diversity and affirming the need for the full participation of women at all levels of policymaking and implementation for biological diversity conservation,

Stressing the importance of, and the need to promote, international, regional and global cooperation among States and intergovernmental organizations and the non-governmental sector for the conservation of biological diversity and the sustainable use of its components,

Acknowledging that the provision of new and additional financial resources and appropriate access to relevant technologies can be expected to make a substantial difference in the world's ability to address the loss of biological diversity,

Acknowledging further that special provision is required to meet the needs of developing countries, including the provision of new and additional financial resources and appropriate access to relevant technologies,

Noting in this regard the special conditions of the least developed countries and small island States,

Acknowledging that substantial investments are required to conserve biological diversity and that there is the expectation of a broad range of environmental, economic and social benefits from those investments,

Recognizing that economic and social development and poverty eradication are the first and overriding priorities of developing countries,

Aware that conservation and sustainable use of biological diversity is of critical importance for meeting the food, health and other needs of the growing world population, for which purpose access to and sharing of both genetic resources and technologies are essential,

Noting that, ultimately, the conservation and sustainable use of biological diversity will strengthen friendly relations among States and contribute to peace for humankind,

Desiring to enhance and complement existing international arrangements for the conservation of biological diversity and sustainable use of its components, and

Determined to conserve and sustainably use biological diversity for the benefit of present and future generations,

Have agreed as follows:

Article 1
Objectives

The objectives of this Convention, to be pursued in accordance with its relevant provisions, are the conservation of biological diversity, the sustainable use of its components and the fair and equitable sharing of the benefits arising out of the utilization of genetic resources, including by appropriate access to genetic resources and by appropriate transfer of relevant technologies, taking into account all rights over those resources and to technologies, and by appropriate funding.

Article 2
Use of Terms

For the purposes of this Convention:

"Biological diversity" means the variability among living organisms from all sources including, inter alia, terrestrial, marine and other aquatic ecosystems and the ecological complexes of which they are part; this includes diversity within species, between species and of ecosystems.

"Biological resources" includes genetic resources, organisms or parts thereof, populations, or any other biotic component of ecosystems with actual or potential use or value for humanity.

"Biotechnology" means any technological application that uses biological systems, living organisms, or derivatives thereof, to make or modify products or processes for specific use.

"Country of origin of genetic resources" means the country which possesses those genetic resources in *in-situ* conditions.

"Country providing genetic resources" means the country supplying genetic resources collected from *in-situ* sources, including populations of both wild and domesticated species, or taken from *ex-situ* sources, which may or may not have originated in that country.

"Domesticated or cultivated species" means species in which the evolutionary process has been influenced by humans to meet their needs.

"Ecosystem" means a dynamic complex of plant, animal and micro-organism communities and their non-living environment interacting as a functional unit.

"*Ex-situ* conservation" means the conservation of components of biological diversity outside their natural habitats.

"Genetic material" means any material of plant, animal, microbial or other origin containing functional units of heredity.

"Genetic resources" means genetic material of actual or potential value.

"Habitat" means the place or type of site where an organism or population naturally occurs.

"*In-situ* conditions" means conditions where genetic resources exist within ecosystems and natural habitats, and, in the case of domesticated or cultivated species, in the surroundings where they have developed their distinctive properties.

"*In-situ* conservation" means the conservation of ecosystems and natural habitats and the maintenance and recovery of viable populations of species in their natural surroundings and, in the case of domesticated or cultivated species, in the surroundings where they have developed their distinctive properties.

"Protected area" means a geographically defined area which is designated or regulated and managed to achieve specific conservation objectives.

"Regional economic integration organization" means an organization constituted by sovereign States of a given region, to which its member States have transferred competence in respect of matters governed by this Convention and which has been duly authorized, in accordance with its internal procedures, to sign, ratify, accept, approve or accede to it.

"Sustainable use" means the use of components of biological diversity in a way and at a rate that does not lead to the long-term decline of biological diversity, thereby maintaining its potential to meet the needs and aspirations of present and future generations.

"Technology" includes biotechnology.

Article 3
Principle

States have, in accordance with the Charter of the United Nations and the principles of international law, the sovereign right to exploit their own resources pursuant to their own environmental policies, and the responsibility to ensure that activities within their jurisdiction or control do not cause damage to the environment of other States or of areas beyond the limits of national jurisdiction.

Article 4
Jurisdictional Scope

Subject to the rights of other States, and except as otherwise expressly provided in this Convention, the provisions of this Convention apply, in relation to each Contracting Party:

(a) In the case of components of biological diversity, in areas within the limits of its national jurisdiction; and
(b) In the case of processes and activities, regardless of where their effects occur, carried out under its jurisdiction or control, within the area of its national jurisdiction or beyond the limits of national jurisdiction.

Article 5
Cooperation

Each Contracting Party shall, as far as possible and as appropriate, cooperate with other Contracting Parties, directly or, where appropriate, through competent international organizations, in respect of areas beyond national jurisdiction and on other matters of mutual interest, for the conservation and sustainable use of biological diversity.

Article 6
General Measures for Conservation and Sustainable Use

Each Contracting Party shall, in accordance with its particular conditions and capabilities:

(a) Develop national strategies, plans or programmes for the conservation and sustainable use of biological diversity or adapt for this purpose existing strategies, plans or programmes which shall reflect, inter alia, the measures set out in this Convention relevant to the Contracting Party concerned; and
(b) Integrate, as far as possible and as appropriate, the conservation and sustainable use of biological diversity into relevant sectoral or cross-sectoral plans, programmes and policies.

Article 7
Identification and Monitoring

Each Contracting Party shall, as far as possible and as appropriate, in particular for the purposes of Articles 8 to 10:

(a) Identify components of biological diversity important for its conservation and sustainable use having regard to the indicative list of categories set down in Annex I;

(b) Monitor, through sampling and other techniques, the components of biological diversity identified pursuant to subparagraph (a) above, paying particular attention to those requiring urgent conservation measures and those which offer the greatest potential for sustainable use;

(c) Identify processes and categories of activities which have or are likely to have significant adverse impacts on the conservation and sustainable use of biological diversity, and monitor their effects through sampling and other techniques; and

(d) Maintain and organize, by any mechanism data, derived from identification and monitoring activities pursuant to subparagraphs (a), (b) and (c) above.

Article 8
In-situ Conservation

Each Contracting Party shall, as far as possible and as appropriate:

(a) Establish a system of protected areas or areas where special measures need to be taken to conserve biological diversity;

(b) Develop, where necessary, guidelines for the selection, establishment and management of protected areas or areas where special measures need to be taken to conserve biological diversity;

(c) Regulate or manage biological resources important for the conservation of biological diversity whether within or outside protected areas, with a view to ensuring their conservation and sustainable use;
(d) Promote the protection of ecosystems, natural habitats and the maintenance of viable populations of species in natural surroundings;
(e) Promote environmentally sound and sustainable development in areas adjacent to protected areas with a view to furthering protection of these areas;
(f) Rehabilitate and restore degraded ecosystems and promote the recovery of threatened species, inter alia, through the development and implementation of plans or other management strategies;
(g) Establish or maintain means to regulate, manage or control the risks associated with the use and release of living modified organisms resulting from biotechnology which are likely to have adverse environmental impacts that could affect the conservation and sustainable use of biological diversity, taking also into account the risks to human health;
(h) Prevent the introduction of, control or eradicate those alien species which threaten ecosystems, habitats or species;
(i) Endeavour to provide the conditions needed for compatibility between present uses and the conservation of biological diversity and the sustainable use of its components;
(j) Subject to its national legislation, respect, preserve and maintain knowledge, innovations and practices of indigenous and local communities embodying traditional lifestyles relevant for the conservation and sustainable use of biological diversity and promote their wider application with the approval and involvement of the holders of such knowledge, innovations and practices and encourage the equitable sharing of the benefits arising from the utilization of such

knowledge, innovations and practices;

(k) Develop or maintain necessary legislation and/or other regulatory provisions for the protection of threatened species and populations;

(l) Where a significant adverse effect on biological diversity has been determined pursuant to Article 7, regulate or manage the relevant processes and categories of activities; and

(m) Cooperate in providing financial and other support for *in-situ* conservation outlined in subparagraphs (a) to (l) above, particularly to developing countries.

Article 9
Ex-situ Conservation

Each Contracting Party shall, as far as possible and as appropriate, and predominantly for the purpose of complementing *in-situ* measures:

(a) Adopt measures for the *ex-situ* conservation of components of biological diversity, preferably in the country of origin of such components;

(b) Establish and maintain facilities for *ex-situ* conservation of and research on plants, animals and micro-organisms, preferably in the country of origin of genetic resources;

(c) Adopt measures for the recovery and rehabilitation of threatened species and for their reintroduction into their natural habitats under appropriate conditions;

(d) Regulate and manage collection of biological resources from natural habitats for *ex-situ* conservation purposes so as not to threaten ecosystems and *in-situ* populations of species, except where special temporary *ex-situ* measures are required under subparagraph (c) above; and

(e) Cooperate in providing financial and other support for exsitu conservation outlined in subparagraphs (a) to (d) above and in the establishment and maintenance of *ex-situ* conservation facilities in developing countries.

Article 10
Sustainable Use of Components of Biological Diversity

Each Contracting Party shall, as far as possible and as appropriate:

(a) Integrate consideration of the conservation and sustainable use of biological resources into national decision-making;
(b) Adopt measures relating to the use of biological resources to avoid or minimize adverse impacts on biological diversity;
(c) Protect and encourage customary use of biological resources in accordance with traditional cultural practices that are compatible with conservation or sustainable use requirements;
(d) Support local populations to develop and implement remedial action in degraded areas where biological diversity has been reduced; and
(e) Encourage cooperation between its governmental authorities and its private sector in developing methods for sustainable use of biological resources.

Article 11
Incentive Measures

Each Contracting Party shall, as far as possible and as appropriate, adopt economically and socially sound measures that act as incentives for the conservation and sustainable use of components of biological diversity.

Article 12
Research and Training

The Contracting Parties, taking into account the special needs of developing countries, shall:

(a) Establish and maintain programmes for scientific and technical education and training in measures for the identification, conservation and sustainable use of biological diversity and its components and provide support for such education and training for the specific needs of developing countries;
(b) Promote and encourage research which contributes to the conservation and sustainable use of biological diversity, particularly in developing countries, inter alia, in accordance with decisions of the Conference of the Parties taken in consequence of recommendations of the Subsidiary Body on Scientific, Technical and Technological Advice; and
(c) In keeping with the provisions of Articles 16, 18 and 20, promote and cooperate in the use of scientific advances in biological diversity research in developing methods for conservation and sustainable use of biological resources.

Article 13
Public Education and Awareness

The Contracting Parties shall:

(a) Promote and encourage understanding of the importance of, and the measures required for, the conservation of biological diversity, as well as its propagation through media, and the inclusion of these topics in educational programmes; and
(b) Cooperate, as appropriate, with other States and international organizations in developing educational and public awareness programmes, with respect to conservation and

sustainable use of biological diversity.

Article 14
Impact Assessment and Minimizing Adverse Impacts

1. Each Contracting Party, as far as possible and as appropriate, shall:

(a) Introduce appropriate procedures requiring environmental impact assessment of its proposed projects that are likely to have significant adverse effects on biological diversity with a view to avoiding or minimizing such effects and, where appropriate, allow for public participation in such procedures;
(b) Introduce appropriate arrangements to ensure that the environmental consequences of its programmes and policies that are likely to have significant adverse impacts on biological diversity are duly taken into account;
(c) Promote, on the basis of reciprocity, notification, exchange of information and consultation on activities under their jurisdiction or control which are likely to significantly affect adversely the biological diversity of other States or areas beyond the limits of national jurisdiction, by encouraging the conclusion of bilateral, regional or multilateral arrangements, as appropriate;
(d) In the case of imminent or grave danger or damage, originating under its jurisdiction or control, to biological diversity within the area under jurisdiction of other States or in areas beyond the limits of national jurisdiction, notify immediately the potentially affected States of such danger or damage, as well as initiate action to prevent or minimize such danger or damage; and
(e) Promote national arrangements for emergency responses to activities or events, whether caused naturally or otherwise, which present a grave and imminent danger to biological

diversity and encourage international cooperation to supplement such national efforts and, where appropriate and agreed by the States or regional economic integration organizations concerned, to establish joint contingency plans.

2. The Conference of the Parties shall examine, on the basis of studies to be carried out, the issue of liability and redress, including restoration and compensation, for damage to biological diversity, except where such liability is a purely internal matter.

Article 15
Access to Genetic Resources

1. Recognizing the sovereign rights of States over their natural resources, the authority to determine access to genetic resources rests with the national governments and is subject to national legislation.

2. Each Contracting Party shall endeavour to create conditions to facilitate access to genetic resources for environmentally sound uses by other Contracting Parties and not to impose restrictions that run counter to the objectives of this Convention.

3. For the purpose of this Convention, the genetic resources being provided by a Contracting Party, as referred to in this Article and Articles 16 and 19, are only those that are provided by Contracting Parties that are countries of origin of such resources or by the Parties that have acquired the genetic resources in accordance with this Convention.

4. Access, where granted, shall be on mutually agreed terms and subject to the provisions of this Article.

5. Access to genetic resources shall be subject to prior informed consent of the Contracting Party providing such resources, unless otherwise determined by that Party.

6. Each Contracting Party shall endeavour to develop and carry out scientific research based on genetic resources provided by other Contracting Parties with the full participation of, and where possible in, such Contracting Parties.

7. Each Contracting Party shall take legislative, administrative or policy measures, as appropriate, and in accordance with Articles 16 and 19 and, where necessary, through the financial mechanism established by Articles 20 and 21 with the aim of sharing in a fair and equitable way the results of research and development and the benefits arising from the commercial and other utilization of genetic resources with the Contracting Party providing such resources. Such sharing shall be upon mutually agreed terms.

Article 16
Access to and Transfer of Technology

1. Each Contracting Party, recognizing that technology includes biotechnology, and that both access to and transfer of technology among Contracting Parties are essential elements for the attainment of the objectives of this Convention, undertakes subject to the provisions of this Article to provide and/or facilitate access for and transfer to other Contracting Parties of technologies that are relevant to the conservation and sustainable use of biological diversity or make use of genetic resources and do not cause significant damage to the environment.

2. Access to and transfer of technology referred to in paragraph 1 above to developing countries shall be provided and/or facilitated under fair and most favourable terms, including on concessional and preferential terms where mutually agreed, and, where necessary, in accordance with the financial mechanism established by Articles 20 and 21. In the case of technology subject to patents and other intellectual property rights, such access and transfer shall be provided on terms which recognize and are

consistent with the adequate and effective protection of intellectual property rights. The application of this paragraph shall be consistent with paragraphs 3, 4 and 5 below.

3. Each Contracting Party shall take legislative, administrative or policy measures, as appropriate, with the aim that Contracting Parties, in particular those that are developing countries, which provide genetic resources are provided access to and transfer of technology which makes use of those resources, on mutually agreed terms, including technology protected by patents and other intellectual property rights, where necessary, through the provisions of Articles 20 and 21 and in accordance with international law and consistent with paragraphs 4 and 5 below.

4. Each Contracting Party shall take legislative, administrative or policy measures, as appropriate, with the aim that the private sector facilitates access to, joint development and transfer of technology referred to in paragraph 1 above for the benefit of both governmental institutions and the private sector of developing countries and in this regard shall abide by the obligations included in paragraphs 1, 2 and 3 above.

5. The Contracting Parties, recognizing that patents and other intellectual property rights may have an influence on the implementation of this Convention, shall cooperate in this regard subject to national legislation and international law in order to ensure that such rights are supportive of and do not run counter to its objectives.

Article 17
Exchange of Information

1. The Contracting Parties shall facilitate the exchange of information, from all publicly available sources, relevant to the conservation and sustainable use of biological diversity, taking into

account the special needs of developing countries.

2. Such exchange of information shall include exchange of results of technical, scientific and socio-economic research, as well as information on training and surveying programmes, specialized knowledge, indigenous and traditional knowledge as such and in combination with the technologies referred to in Article 16, paragraph 1. It shall also, where feasible, include repatriation of information.

Article 18
Technical and Scientific Cooperation

1. The Contracting Parties shall promote international technical and scientific cooperation in the field of conservation and sustainable use of biological diversity, where necessary, through the appropriate international and national institutions.

2. Each Contracting Party shall promote technical and scientific cooperation with other Contracting Parties, in particular developing countries, in implementing this Convention, inter alia, through the development and implementation of national policies. In promoting such cooperation, special attention should be given to the development and strengthening of national capabilities, by means of human resources development and institution building.

3. The Conference of the Parties, at its first meeting, shall determine how to establish a clearing-house mechanism to promote and facilitate technical and scientific cooperation.

4. The Contracting Parties shall, in accordance with national legislation and policies, encourage and develop methods of cooperation for the development and use of technologies, including indigenous and traditional technologies, in pursuance of the objectives of this Convention. For this purpose, the Contracting Parties shall also promote cooperation in the training of personnel

and exchange of experts.

5. The Contracting Parties shall, subject to mutual agreement, promote the establishment of joint research programmes and joint ventures for the development of technologies relevant to the objectives of this Convention.

Article 19
Handling of Biotechnology and Distribution of its Benefits

1. Each Contracting Party shall take legislative, administrative or policy measures, as appropriate, to provide for the effective participation in biotechnological research activities by those Contracting Parties, especially developing countries, which provide the genetic resources for such research, and where feasible in such Contracting Parties.

2. Each Contracting Party shall take all practicable measures to promote and advance priority access on a fair and equitable basis by Contracting Parties, especially developing countries, to the results and benefits arising from biotechnologies based upon genetic resources provided by those Contracting Parties. Such access shall be on mutually agreed terms.

3. The Parties shall consider the need for and modalities of a protocol setting out appropriate procedures, including, in particular, advance informed agreement, in the field of the safe transfer, handling and use of any living modified organism resulting from biotechnology that may have adverse effect on the conservation and sustainable use of biological diversity.

4. Each Contracting Party shall, directly or by requiring any natural or legal person under its jurisdiction providing the organisms referred to in paragraph 3 above, provide any available information about the use and safety regulations required by that Contracting Party in handling such organisms, as well as any

available information on the potential adverse impact of the specific organisms concerned to the Contracting Party into which those organisms are to be introduced.

Article 20
Financial Resources

1. Each Contracting Party undertakes to provide, in accordance with its capabilities, financial support and incentives in respect of those national activities which are intended to achieve the objectives of this Convention, in accordance with its national plans, priorities and programmes.

2. The developed country Parties shall provide new and additional financial resources to enable developing country Parties to meet the agreed full incremental costs to them of implementing measures which fulfil the obligations of this Convention and to benefit from its provisions and which costs are agreed between a developing country Party and the institutional structure referred to in Article 21, in accordance with policy, strategy, programme priorities and eligibility criteria and an indicative list of incremental costs established by the Conference of the Parties. Other Parties, including countries undergoing the process of transition to a market economy, may voluntarily assume the obligations of the developed country Parties. For the purpose of this Article, the Conference of the Parties, shall at its first meeting establish a list of developed country Parties and other Parties which voluntarily assume the obligations of the developed country Parties. The Conference of the Parties shall periodically review and if necessary amend the list. Contributions from other countries and sources on a voluntary basis would also be encouraged. The implementation of these commitments shall take into account the need for adequacy, predictability and timely flow of funds and the importance of burden-sharing among the

contributing Parties included in the list.

3. The developed country Parties may also provide, and developing country Parties avail themselves of, financial resources related to the implementation of this Convention through bilateral, regional and other multilateral channels.

4. The extent to which developing country Parties will effectively implement their commitments under this Convention will depend on the effective implementation by developed country Parties of their commitments under this Convention related to financial resources and transfer of technology and will take fully into account the fact that economic and social development and eradication of poverty are the first and overriding priorities of the developing country Parties.

5. The Parties shall take full account of the specific needs and special situation of least developed countries in their actions with regard to funding and transfer of technology.

6. The Contracting Parties shall also take into consideration the special conditions resulting from the dependence on, distribution and location of, biological diversity within developing country Parties, in particular small island States.

7. Consideration shall also be given to the special situation of developing countries, including those that are most environmentally vulnerable, such as those with arid and semi-arid zones, coastal and mountainous areas.

Article 21
Financial Mechanism

1. There shall be a mechanism for the provision of financial resources to developing country Parties for purposes of this Convention on a grant or concessional basis the essential elements

of which are described in this Article. The mechanism shall function under the authority and guidance of, and be accountable to, the Conference of the Parties for purposes of this Convention. The operations of the mechanism shall be carried out by such institutional structure as may be decided upon by the Conference of the Parties at its first meeting. For purposes of this Convention, the Conference of the Parties shall determine the policy, strategy, programme priorities and eligibility criteria relating to the access to and utilization of such resources. The contributions shall be such as to take into account the need for predictability, adequacy and timely flow of funds referred to in Article 20 in accordance with the amount of resources needed to be decided periodically by the Conference of the Parties and the importance of burden-sharing among the contributing Parties included in the list referred to in Article 20, paragraph 2. Voluntary contributions may also be made by the developed country Parties and by other countries and sources. The mechanism shall operate within a democratic and transparent system of governance.

2. Pursuant to the objectives of this Convention, the Conference of the Parties shall at its first meeting determine the policy, strategy and programme priorities, as well as detailed criteria and guidelines for eligibility for access to and utilization of the financial resources including monitoring and evaluation on a regular basis of such utilization. The Conference of the Parties shall decide on the arrangements to give effect to paragraph 1 above after consultation with the institutional structure entrusted with the operation of the financial mechanism.

3. The Conference of the Parties shall review the effectiveness of the mechanism established under this Article, including the criteria and guidelines referred to in paragraph 2 above, not less than two years after the entry into force of this Convention and thereafter on a regular basis. Based on such review, it shall take

appropriate action to improve the effectiveness of the mechanism if necessary.

4. The Contracting Parties shall consider strengthening existing financial institutions to provide financial resources for the conservation and sustainable use of biological diversity.

Article 22
Relationship with Other International Conventions

1. The provisions of this Convention shall not affect the rights and obligations of any Contracting Party deriving from any existing international agreement, except where the exercise of those rights and obligations would cause a serious damage or threat to biological diversity.

2. Contracting Parties shall implement this Convention with respect to the marine environment consistently with the rights and obligations of States under the law of the sea.

Article 23
Conference of the Parties

1. A Conference of the Parties is hereby established. The first meeting of the Conference of the Parties shall be convened by the Executive Director of the United Nations Environment Programme not later than one year after the entry into force of this Convention. Thereafter, ordinary meetings of the Conference of the Parties shall be held at regular intervals to be determined by the Conference at its first meeting.

2. Extraordinary meetings of the Conference of the Parties shall be held at such other times as may be deemed necessary by the Conference, or at the written request of any Party, provided that, within six months of the request being communicated to them by

the Secretariat, it is supported by at least one third of the Parties.

3. The Conference of the Parties shall by consensus agree upon and adopt rules of procedure for itself and for any subsidiary body it may establish, as well as financial rules governing the funding of the Secretariat. At each ordinary meeting, it shall adopt a budget for the financial period until the next ordinary meeting.

4. The Conference of the Parties shall keep under review the implementation of this Convention, and, for this purpose, shall:

(a) Establish the form and the intervals for transmitting the information to be submitted in accordance with Article 26 and consider such information as well as reports submitted by any subsidiary body;
(b) Review scientific, technical and technological advice on biological diversity provided in accordance with Article 25;
(c) Consider and adopt, as required, protocols in accordance with Article 28;
(d) Consider and adopt, as required, in accordance with Articles 29 and 30, amendments to this Convention and its annexes;
(e) Consider amendments to any protocol, as well as to any annexes thereto, and, if so decided, recommend their adoption to the parties to the protocol concerned;
(f) Consider and adopt, as required, in accordance with Article 30, additional annexes to this Convention;
(g) Establish such subsidiary bodies, particularly to provide scientific and technical advice, as are deemed necessary for the implementation of this Convention;
(h) Contact, through the Secretariat, the executive bodies of conventions dealing with matters covered by this Convention with a view to establishing appropriate forms of cooperation with them; and

(i) Consider and undertake any additional action that may be required for the achievement of the purposes of this Convention in the light of experience gained in its operation.

5. The United Nations, its specialized agencies and the International Atomic Energy Agency, as well as any State not Party to this Convention, may be represented as observers at meetings of the Conference of the Parties. Any other body or agency, whether governmental or non-governmental, qualified in fields relating to conservation and sustainable use of biological diversity, which has informed the Secretariat of its wish to be represented as an observer at a meeting of the Conference of the Parties, may be admitted unless at least one third of the Parties present object. The admission and participation of observers shall be subject to the rules of procedure adopted by the Conference of the Parties.

Article 24
Secretariat

1. A secretariat is hereby established. Its functions shall be:

(a) To arrange for and service meetings of the Conference of the Parties provided for in Article 23;
(b) To perform the functions assigned to it by any protocol;
(c) To prepare reports on the execution of its functions under this Convention and present them to the Conference of the Parties;
(d) To coordinate with other relevant international bodies and, in particular to enter into such administrative and contractual arrangements as may be required for the effective discharge of its functions; and
(e) To perform such other functions as may be determined by the Conference of the Parties.

2. At its first ordinary meeting, the Conference of the Parties shall designate the secretariat from amongst those existing competent international organizations which have signified their willingness to carry out the secretariat functions under this Convention.

Article 25
Subsidiary Body on Scientific, Technical and Technological Advice

1. A subsidiary body for the provision of scientific, technical and technological advice is hereby established to provide the Conference of the Parties and, as appropriate, its other subsidiary bodies with timely advice relating to the implementation of this Convention. This body shall be open to participation by all Parties and shall be multidisciplinary. It shall comprise government representatives competent in the relevant field of expertise. It shall report regularly to the Conference of the Parties on all aspects of its work.

2. Under the authority of and in accordance with guidelines laid down by the Conference of the Parties, and upon its request, this body shall:

(a) Provide scientific and technical assessments of the status of biological diversity;
(b) Prepare scientific and technical assessments of the effects of types of measures taken in accordance with the provisions of this Convention;
(c) Identify innovative, efficient and state-of-the-art technologies and know-how relating to the conservation and sustainable use of biological diversity and advise on the ways and means of promoting development and/or transferring such technologies;
(d) Provide advice on scientific programmes and international cooperation in research and development related to

conservation and sustainable use of biological diversity; and

(e) Respond to scientific, technical, technological and methodological questions that the Conference of the Parties and its subsidiary bodies may put to the body.

3. The functions, terms of reference, organization and operation of this body may be further elaborated by the Conference of the Parties.

Article 26
Reports

Each Contracting Party shall, at intervals to be determined by the Conference of the Parties, present to the Conference of the Parties, reports on measures which it has taken for the implementation of the provisions of this Convention and their effectiveness in meeting the objectives of this Convention.

Article 27
Settlement of Disputes

1. In the event of a dispute between Contracting Parties concerning the interpretation or application of this Convention, the parties concerned shall seek solution by negotiation.

2. If the parties concerned cannot reach agreement by negotiation, they may jointly seek the good offices of, or request mediation by, a third party.

3. When ratifying, accepting, approving or acceding to this Convention, or at any time thereafter, a State or regional economic integration organization may declare in writing to the Depositary that for a dispute not resolved in accordance with paragraph 1 or paragraph 2 above, it accepts one or both of the following means of dispute settlement as compulsory:

(a) Arbitration in accordance with the procedure laid down in Part 1 of Annex II;
(b) Submission of the dispute to the International Court of Justice.

4. If the parties to the dispute have not, in accordance with paragraph 3 above, accepted the same or any procedure, the dispute shall be submitted to conciliation in accordance with Part 2 of Annex II unless the parties otherwise agree.

5. The provisions of this Article shall apply with respect to any protocol except as otherwise provided in the protocol concerned.

Article 28
Adoption of Protocols

1. The Contracting Parties shall cooperate in the formulation and adoption of protocols to this Convention.

2. Protocols shall be adopted at a meeting of the Conference of the Parties.

3. The text of any proposed protocol shall be communicated to the Contracting Parties by the Secretariat at least six months before such a meeting.

Article 29
Amendment of the Convention or Protocols

1. Amendments to this Convention may be proposed by any Contracting Party. Amendments to any protocol may be proposed by any Party to that protocol.

2. Amendments to this Convention shall be adopted at a meeting of the Conference of the Parties. Amendments to any protocol shall be adopted at a meeting of the Parties to the Protocol in

question. The text of any proposed amendment to this Convention or to any protocol, except as may otherwise be provided in such protocol, shall be communicated to the Parties to the instrument in question by the secretariat at least six months before the meeting at which it is proposed for adoption. The secretariat shall also communicate proposed amendments to the signatories to this Convention for information.

3. The Parties shall make every effort to reach agreement on any proposed amendment to this Convention or to any protocol by consensus. If all efforts at consensus have been exhausted, and no agreement reached, the amendment shall as a last resort be adopted by a two-third majority vote of the Parties to the instrument in question present and voting at the meeting, and shall be submitted by the Depositary to all Parties for ratification, acceptance or approval.

4. Ratification, acceptance or approval of amendments shall be notified to the Depositary in writing. Amendments adopted in accordance with paragraph 3 above shall enter into force among Parties having accepted them on the ninetieth day after the deposit of instruments of ratification, acceptance or approval by at least two thirds of the Contracting Parties to this Convention or of the Parties to the protocol concerned, except as may otherwise be provided in such protocol. Thereafter the amendments shall enter into force for any other Party on the ninetieth day after that Party deposits its instrument of ratification, acceptance or approval of the amendments.

5. For the purposes of this Article, “Parties present and voting” means Parties present and casting an affirmative or negative vote.

Article 30
Adoption and Amendment of Annexes

1. The annexes to this Convention or to any protocol shall form an integral part of the Convention or of such protocol, as the case may be, and, unless expressly provided otherwise, a reference to this Convention or its protocols constitutes at the same time a reference to any annexes thereto. Such annexes shall be restricted to procedural, scientific, technical and administrative matters.

2. Except as may be otherwise provided in any protocol with respect to its annexes, the following procedure shall apply to the proposal, adoption and entry into force of additional annexes to this Convention or of annexes to any protocol:

(a) Annexes to this Convention or to any protocol shall be proposed and adopted according to the procedure laid down in Article 29;

(b) Any Party that is unable to approve an additional annex to this Convention or an annex to any protocol to which it is Party shall so notify the Depositary, in writing, within one year from the date of the communication of the adoption by the Depositary. The Depositary shall without delay notify all Parties of any such notification received. A Party may at any time withdraw a previous declaration of objection and the annexes shall thereupon enter into force for that Party subject to subparagraph (c) below;

(c) On the expiry of one year from the date of the communication of the adoption by the Depositary, the annex shall enter into force for all Parties to this Convention or to any protocol concerned which have not submitted a notification in accordance with the provisions of subparagraph (b) above.

3. The proposal, adoption and entry into force of amendments to

annexes to this Convention or to any protocol shall be subject to the same procedure as for the proposal, adoption and entry into force of annexes to the Convention or annexes to any protocol.

4. If an additional annex or an amendment to an annex is related to an amendment to this Convention or to any protocol, the additional annex or amendment shall not enter into force until such time as the amendment to the Convention or to the protocol concerned enters into force.

Article 31
Right to Vote

1. Except as provided for in paragraph 2 below, each Contracting Party to this Convention or to any protocol shall have one vote.

2. Regional economic integration organizations, in matters within their competence, shall exercise their right to vote with a number of votes equal to the number of their member States which are Contracting Parties to this Convention or the relevant protocol. Such organizations shall not exercise their right to vote if their member States exercise theirs, and vice versa.

Article 32
Relationship between this Convention and Its Protocols

1. A State or a regional economic integration organization may not become a Party to a protocol unless it is, or becomes at the same time, a Contracting Party to this Convention.

2. Decisions under any protocol shall be taken only by the Parties to the protocol concerned. Any Contracting Party that has not ratified, accepted or approved a protocol may participate as an observer in any meeting of the parties to that protocol.

Article 33
Signature

This Convention shall be open for signature at Rio de Janeiro by all States and any regional economic integration organization from 5 June 1992 until 14 June 1992, and at the United Nations Headquarters in New York from 15 June 1992 to 4 June 1993.

Article 34
Ratification, Acceptance or Approval

1. This Convention and any protocol shall be subject to ratification, acceptance or approval by States and by regional economic integration organizations. Instruments of ratification, acceptance or approval shall be deposited with the Depositary.

2. Any organization referred to in paragraph 1 above which becomes a Contracting Party to this Convention or any protocol without any of its member States being a Contracting Party shall be bound by all the obligations under the Convention or the protocol, as the case may be. In the case of such organizations, one or more of whose member States is a Contracting Party to this Convention or relevant protocol, the organization and its member States shall decide on their respective responsibilities for the performance of their obligations under the Convention or protocol, as the case may be. In such cases, the organization and the member States shall not be entitled to exercise rights under the Convention or relevant protocol concurrently.

3. In their instruments of ratification, acceptance or approval, the organizations referred to in paragraph 1 above shall declare the extent of their competence with respect to the matters governed by the Convention or the relevant protocol. These organizations shall also inform the Depositary of any relevant modification in

the extent of their competence.

Article 35
Accession

1. This Convention and any protocol shall be open for accession by States and by regional economic integration organizations from the date on which the Convention or the protocol concerned is closed for signature. The instruments of accession shall be deposited with the Depositary.

2. In their instruments of accession, the organizations referred to in paragraph 1 above shall declare the extent of their competence with respect to the matters governed by the Convention or the relevant protocol. These organizations shall also inform the Depositary of any relevant modification in the extent of their competence.

3. The provisions of Article 34, paragraph 2, shall apply to regional economic integration organizations which accede to this Convention or any protocol.

Article 36
Entry Into Force

1. This Convention shall enter into force on the ninetieth day after the date of deposit of the thirtieth instrument of ratification, acceptance, approval or accession.

2. Any protocol shall enter into force on the ninetieth day after the date of deposit of the number of instruments of ratification, acceptance, approval or accession, specified in that protocol, has been deposited.

3. For each Contracting Party which ratifies, accepts or approves this Convention or accedes thereto after the deposit of the

thirtieth instrument of ratification, acceptance, approval or accession, it shall enter into force on the ninetieth day after the date of deposit by such Contracting Party of its instrument of ratification, acceptance, approval or accession.

4. Any protocol, except as otherwise provided in such protocol, shall enter into force for a Contracting Party that ratifies, accepts or approves that protocol or accedes thereto after its entry into force pursuant to paragraph 2 above, on the ninetieth day after the date on which that Contracting Party deposits its instrument of ratification, acceptance, approval or accession, or on the date on which this Convention enters into force for that Contracting Party, whichever shall be the later.

5. For the purposes of paragraphs 1 and 2 above, any instrument deposited by a regional economic integration organization shall not be counted as additional to those deposited by member States of such organization.

Article 37
Reservations

No reservations may be made to this Convention.

Article 38
Withdrawals

1. At any time after two years from the date on which this Convention has entered into force for a Contracting Party, that Contracting Party may withdraw from the Convention by giving written notification to the Depositary.

2. Any such withdrawal shall take place upon expiry of one year after the date of its receipt by the Depositary, or on such later date as may be specified in the notification of the withdrawal.

3. Any Contracting Party which withdraws from this Convention shall be considered as also having withdrawn from any protocol to which it is party.

Article 39
Financial Interim Arrangements

Provided that it has been fully restructured in accordance with the requirements of Article 21, the Global Environment Facility of the United Nations Development Programme, the United Nations Environment Programme and the International Bank for Reconstruction and Development shall be the institutional structure referred to in Article 21 on an interim basis, for the period between the entry into force of this Convention and the first meeting of the Conference of the Parties or until the Conference of the Parties decides which institutional structure will be designated in accordance with Article 21.

Article 40
Secretariat Interim Arrangements

The secretariat to be provided by the Executive Director of the United Nations Environment Programme shall be the secretariat referred to in Article 24, paragraph 2, on an interim basis for the period between the entry into force of this Convention and the first meeting of the Conference of the Parties.

Article 41
Depositary

The Secretary-General of the United Nations shall assume the functions of Depositary of this Convention and any protocols.

Article 42
Authentic Texts

The original of this Convention, of which the Arabic, Chinese, English, French, Russian and Spanish texts are equally authentic, shall be deposited with the Secretary-General of the United Nations.

IN WITNESS WHEREOF the undersigned, being duly authorized to that effect, have signed this Convention.

Done at Rio de Janeiro on this fifth day of June, one thousand nine hundred and ninety-two.

Annex I
Identification and Monitoring

1. Ecosystems and habitats: containing high diversity, large numbers of endemic or threatened species, or wilderness; required by migratory species; of social, economic, cultural or scientific importance; or, which are representative, unique or associated with key evolutionary or other biological processes;

2. Species and communities which are: threatened; wild relatives of domesticated or cultivated species; of medicinal, agricultural or other economic value; or social, scientific or cultural importance; or importance for research into the conservation and sustainable use of biological diversity, such as indicator species; and

3. Described genomes and genes of social, scientific or economic importance.

Annex II

Part 1
ARBITRATION

Article 1

The claimant party shall notify the secretariat that the parties are referring a dispute to arbitration pursuant to Article 27. The notification shall state the subject-matter of arbitration and include, in particular, the articles of the Convention or the protocol, the interpretation or application of which are at issue. If the parties do not agree on the subject matter of the dispute before the President of the tribunal is designated, the arbitral tribunal shall determine the subject matter. The secretariat shall forward the information thus received to all Contracting Parties to this Convention or to the protocol concerned.

Article 2

1. In disputes between two parties, the arbitral tribunal shall consist of three members. Each of the parties to the dispute shall appoint an arbitrator and the two arbitrators so appointed shall designate by common agreement the third arbitrator who shall be the President of the tribunal. The latter shall not be a national of one of the parties to the dispute, nor have his or her usual place of residence in the territory of one of these parties, nor be employed by any of them, nor have dealt with the case in any other capacity.

2. In disputes between more than two parties, parties in the same interest shall appoint one arbitrator jointly by agreement.

3. Any vacancy shall be filled in the manner prescribed for the initial appointment.

Article 3

1. If the President of the arbitral tribunal has not been designated within two months of the appointment of the second arbitrator, the Secretary-General of the United Nations shall, at the request of a party, designate the President within a further two-month period.

2. If one of the parties to the dispute does not appoint an arbitrator within two months of receipt of the request, the other party may inform the Secretary-General who shall make the designation within a further two-month period.

Article 4

The arbitral tribunal shall render its decisions in accordance with the provisions of this Convention, any protocols concerned, and international law.

Article 5

Unless the parties to the dispute otherwise agree, the arbitral tribunal shall determine its own rules of procedure.

Article 6

The arbitral tribunal may, at the request of one of the parties, recommend essential interim measures of protection.

Article 7

The parties to the dispute shall facilitate the work of the arbitral tribunal and, in particular, using all means at their disposal, shall:

(a) Provide it with all relevant documents, information and

facilities; and

(b) Enable it, when necessary, to call witnesses or experts and receive their evidence.

Article 8

The parties and the arbitrators are under an obligation to protect the confidentiality of any information they receive in confidence during the proceedings of the arbitral tribunal.

Article 9

Unless the arbitral tribunal determines otherwise because of the particular circumstances of the case, the costs of the tribunal shall be borne by the parties to the dispute in equal shares. The tribunal shall keep a record of all its costs, and shall furnish a final statement thereof to the parties.

Article 10

Any Contracting Party that has an interest of a legal nature in the subject-matter of the dispute which may be affected by the decision in the case, may intervene in the proceedings with the consent of the tribunal.

Article 11

The tribunal may hear and determine counterclaims arising directly out of the subject-matter of the dispute.

Article 12

Decisions both on procedure and substance of the arbitral tribunal shall be taken by a majority vote of its members.

Article 13

If one of the parties to the dispute does not appear before the arbitral tribunal or fails to defend its case, the other party may request the tribunal to continue the proceedings and to make its award. Absence of a party or a failure of a party to defend its case shall not constitute a bar to the proceedings. Before rendering its final decision, the arbitral tribunal must satisfy itself that the claim is well founded in fact and law.

Article 14

The tribunal shall render its final decision within five months of the date on which it is fully constituted unless it finds it necessary to extend the time-limit for a period which should not exceed five more months.

Article 15

The final decision of the arbitral tribunal shall be confined to the subject-matter of the dispute and shall state the reasons on which it is based. It shall contain the names of the members who have participated and the date of the final decision. Any member of the tribunal may attach a separate or dissenting opinion to the final decision.

Article 16

The award shall be binding on the parties to the dispute. It shall be without appeal unless the parties to the dispute have agreed in advance to an appellate procedure.

Article 17

Any controversy which may arise between the parties to the dispute as regards the interpretation or manner of implementation of the final decision may be submitted by either party for decision to the arbitral tribunal which rendered it.

Part 2
CONCILIATION

Article 1

A conciliation commission shall be created upon the request of one of the parties to the dispute. The commission shall, unless the parties otherwise agree, be composed of five members, two appointed by each Party concerned and a President chosen jointly by those members.

Article 2

In disputes between more than two parties, parties in the same interest shall appoint their members of the commission jointly by agreement. Where two or more parties have separate interests or there is a disagreement as to whether they are of the same interest, they shall appoint their members separately.

Article 3

If any appointments by the parties are not made within two months of the date of the request to create a conciliation commission, the Secretary-General of the United Nations shall, if asked to do so by the party that made the request, make those appointments within a further two-month period.

Article 4

If a President of the conciliation commission has not been chosen within two months of the last of the members of the commission being appointed, the Secretary-General of the United Nations shall, if asked to do so by a party, designate a President within a further two-month period.

Article 5

The conciliation commission shall take its decisions by majority vote of its members. It shall, unless the parties to the dispute otherwise agree, determine its own procedure. It shall render a proposal for resolution of the dispute, which the parties shall consider in good faith.

Article 6

A disagreement as to whether the conciliation commission has competence shall be decided by the commission.

16. Cartagena Protocol on Biosafety to the Convention on Biological Diversity (2000)

Date: 29 January 2000
In force: 11 September 2003
Contracting Parties: Signatories 103, Parties 159
Korea: 1 January 2008 (No.1877)
Link: www.cbd.int

The Parties to this Protocol,

Being Parties to the Convention on Biological Diversity, hereinafter referred to as "the Convention",

Recalling Article 19, paragraphs 3 and 4, and Articles 8 (g) and 17 of the Convention,

Recalling also decision II/5 of 17 November 1995 of the Conference of the Parties to the Convention to develop a Protocol on biosafety, specifically focusing on transboundary movement of any living modified organism resulting from modern biotechnology that may have adverse effect on the conservation and sustainable use of biological diversity, setting out for consideration, in particular, appropriate procedures for advance informed agreement,

Reaffirming the precautionary approach contained in Principle 15 of the Rio Declaration on Environment and Development,

Aware of the rapid expansion of modern biotechnology and the growing public concern over its potential adverse effects on biological diversity, taking also into account risks to human health,

Recognizing that modern biotechnology has great potential for human wellbeing if developed and used with adequate safety measures for the environment and human health,

Recognizing also the crucial importance to humankind of centres of origin and centres of genetic diversity,

Taking into account the limited capabilities of many countries, particularly developing countries, to cope with the nature and scale of known and potential risks associated with living modified organisms,

Recognizing that trade and environment agreements should be mutually supportive with a view to achieving sustainable development,

Emphasizing that this Protocol shall not be interpreted as implying a change in the rights and obligations of a Party under any existing international agreements,

Understanding that the above recital is not intended to subordinate this Protocol to other international agreements,

Have agreed as follows:

Article 1
Objective

In accordance with the precautionary approach contained in Principle 15 of the Rio Declaration on Environment and Development, the objective of this Protocol is to contribute to ensuring an adequate level of protection in the field of the safe transfer, handling and use of living modified organisms resulting from modern biotechnology that may have adverse effects on the conservation and sustainable use of biological diversity, taking also into account risks to human health, and specifically focusing on transboundary movements.

Article 2
General Provisions

1. Each Party shall take necessary and appropriate legal, administrative and other measures to implement its obligations under this Protocol.

2. The Parties shall ensure that the development, handling, transport, use, transfer and release of any living modified organisms are undertaken in a manner that prevents or reduces the risks to biological diversity, taking also into account risks to human health.

3. Nothing in this Protocol shall affect in any way the sovereignty of States over their territorial sea established in accordance with international law, and the sovereign rights and the jurisdiction which States have in their exclusive economic zones and their continental shelves in accordance with international law, and the exercise by ships and aircraft of all States of navigational rights and freedoms as provided for in international law and as reflected in relevant international instruments.

4. Nothing in this Protocol shall be interpreted as restricting the right of a Party to take action that is more protective of the conservation and sustainable use of biological diversity than that called for in this Protocol, provided that such action is consistent with the objective and the provisions of this Protocol and is in accordance with that Party's other obligations under international law.

5. The Parties are encouraged to take into account, as appropriate, available expertise, instruments and work undertaken in international forums with competence in the area of risks to human health.

Article 3
Use of Terms

For the purposes of this Protocol:

(a) "Conference of the Parties" means the Conference of the Parties to the Convention;
(b) "Contained use" means any operation, undertaken within a facility, installation or other physical structure, which involves living modified organisms that are controlled by specific measures that effectively limit their contact with, and their impact on, the external environment;
(c) "Export" means intentional transboundary movement from one Party to another Party;
(d) "Exporter" means any legal or natural person, under the jurisdiction of the Party of export, who arranges for a living modified organism to be exported;
(e) "Import" means intentional transboundary movement into one Party from another Party;
(f) "Importer" means any legal or natural person, under the jurisdiction of the Party of import, who arranges for a living modified organism to be imported;
(g) "Living modified organism" means any living organism that possesses a novel combination of genetic material obtained through the use of modern biotechnology;
(h) "Living organism" means any biological entity capable of transferring or replicating genetic material, including sterile organisms, viruses and viroids;
(i) "Modern biotechnology" means the application of:
  a. *In vitro* nucleic acid techniques, including recombinant deoxyribonucleic acid(DNA) and direct injection of nucleic acid into cells or organelles, or
  b. Fusion of cells beyond the taxonomic family, that overcome natural physiological reproductive or recombi-

nation barriers and that are not techniques used in traditional breeding and selection;

(j) "Regional economic integration organization" means an organization constituted by sovereign States of a given region, to which its member States have transferred competence in respect of matters governed by this Protocol and which has been duly authorized, in accordance with its internal procedures, to sign, ratify, accept, approve or accede to it;

(k) "Transboundary movement" means the movement of a living modified organism from one Party to another Party, save that for the purposes of Articles 17 and 24 transboundary movement extends to movement between Parties and non-Parties.

Article 4
Scope

This Protocol shall apply to the transboundary movement, transit, handling and use of all living modified organisms that may have adverse effects on the conservation and sustainable use of biological diversity, taking also into account risks to human health.

Article 5
Pharmaceuticals

Notwithstanding Article 4 and without prejudice to any right of a Party to subject all living modified organisms to risk assessment prior to the making of decisions on import, this Protocol shall not apply to the transboundary movement of living modified organisms which are pharmaceuticals for humans that are addressed by other relevant international agreements or organisations.

Article 6
Transit and Contained Use

1. Notwithstanding Article 4 and without prejudice to any right of a Party of transit to regulate the transport of living modified organisms through its territory and make available to the Biosafety Clearing-House, any decision of that Party, subject to Article 2, paragraph 3, regarding the transit through its territory of a specific living modified organism, the provisions of this Protocol with respect to the advance informed agreement procedure shall not apply to living modified organisms in transit.

2. Notwithstanding Article 4 and without prejudice to any right of a Party to subject all living modified organisms to risk assessment prior to decisions on import and to set standards for contained use within its jurisdiction, the provisions of this Protocol with respect to the advance informed agreement procedure shall not apply to the transboundary movement of living modified organisms destined for contained use undertaken in accordance with the standards of the Party of import.

Article 7
Application of the Advance Informed Agreement Procedure

1. Subject to Articles 5 and 6, the advance informed agreement procedure in Articles 8 to 10 and 12 shall apply prior to the first intentional transboundary movement of living modified organisms for intentional introduction into the environment of the Party of import.

2. "Intentional introduction into the environment" in paragraph 1 above, does not refer to living modified organisms intended for direct use as food or feed, or for processing.

3. Article 11 shall apply prior to the first transboundary

movement of living modified organisms intended for direct use as food or feed, or for processing.

4. The advance informed agreement procedure shall not apply to the intentional transboundary movement of living modified organisms identified in a decision of the Conference of the Parties serving as the meeting of the Parties to this Protocol as being not likely to have adverse effects on the conservation and sustainable use of biological diversity, taking also into account risks to human health.

Article 8
Notification

1. The Party of export shall notify, or require the exporter to ensure notification to, in writing, the competent national authority of the Party of import prior to the intentional transboundary movement of a living modified organism that falls within the scope of Article 7, paragraph 1. The notification shall contain, at a minimum, the information specified in Annex I.

2. The Party of export shall ensure that there is a legal requirement for the accuracy of information provided by the exporter.

Article 9
Acknowledgement of Receipt of Notification

1. The Party of import shall acknowledge receipt of the notification, in writing, to the notifier within ninety days of its receipt.

2. The acknowledgement shall state:

(a) The date of receipt of the notification;

(b) Whether the notification, prima facie, contains the information referred to in Article 8;
(c) Whether to proceed according to the domestic regulatory framework of the Party of import or according to the procedure specified in Article 10.

3. The domestic regulatory framework referred to in paragraph 2 (c) above, shall be consistent with this Protocol.

4. A failure by the Party of import to acknowledge receipt of a notification shall not imply its consent to an intentional transboundary movement.

Article 10
Decision Procedure

1. Decisions taken by the Party of import shall be in accordance with Article 15.

2. The Party of import shall, within the period of time referred to in Article 9, inform the notifier, in writing, whether the intentional transboundary movement may proceed:

(a) Only after the Party of import has given its written consent; or
(b) After no less than ninety days without a subsequent written consent.

3. Within two hundred and seventy days of the date of receipt of notification, the Party of import shall communicate, in writing, to the notifier and to the Biosafety Clearing-House the decision referred to in paragraph 2 (a) above:

(a) Approving the import, with or without conditions, including how the decision will apply to subsequent imports of the same living modified organism;
(b) Prohibiting the import;

(c) Requesting additional relevant information in accordance with its domestic regulatory framework or Annex I; in calculating the time within which the Party of import is to respond, the number of days it has to wait for additional relevant information shall not be taken into account; or

(d) Informing the notifier that the period specified in this paragraph is extended by a defined period of time.

4. Except in a case in which consent is unconditional, a decision under paragraph 3 above, shall set out the reasons on which it is based.

5. A failure by the Party of import to communicate its decision within two hundred and seventy days of the date of receipt of the notification shall not imply its consent to an intentional transboundary movement.

6. Lack of scientific certainty due to insufficient relevant scientific information and knowledge regarding the extent of the potential adverse effects of a living modified organism on the conservation and sustainable use of biological diversity in the Party of import, taking also into account risks to human health, shall not prevent that Party from taking a decision, as appropriate, with regard to the import of the living modified organism in question as referred to in paragraph 3 above, in order to avoid or minimize such potential adverse effects.

7. The Conference of the Parties serving as the meeting of the Parties shall, at its first meeting, decide upon appropriate procedures and mechanisms to facilitate decision-making by Parties of import.

Article 11
Procedure for Living Modified Organisms Intended for Direct Use as Food or Feed, or for Processing

1. A Party that makes a final decision regarding domestic use,

including placing on the market, of a living modified organism that may be subject to transboundary movement for direct use as food or feed, or for processing shall, within fifteen days of making that decision, inform the Parties through the Biosafety Clearing-House. This information shall contain, at a minimum, the information specified in Annex II. The Party shall provide a copy of the information, in writing, to the national focal point of each Party that informs the Secretariat in advance that it does not have access to the Biosafety Clearing-House. This provision shall not apply to decisions regarding field trials.

2. The Party making a decision under paragraph 1 above, shall ensure that there is a legal requirement for the accuracy of information provided by the applicant.

3. Any Party may request additional information from the authority identified in paragraph (b) of Annex II.

4. A Party may take a decision on the import of living modified organisms intended for direct use as food or feed, or for processing, under its domestic regulatory framework that is consistent with the objective of this Protocol.

5. Each Party shall make available to the Biosafety Clearing-House copies of any national laws, regulations and guidelines applicable to the import of living modified organisms intended for direct use as food or feed, or for processing, if available.

6. A developing country Party or a Party with an economy in transition may, in the absence of the domestic regulatory framework referred to in paragraph 4 above, and in exercise of its domestic jurisdiction, declare through the Biosafety Clearing-House that its decision prior to the first import of a living modified organism intended for direct use as food or feed, or for processing, on which information has been provided under para-

graph 1 above, will be taken according to the following:

(a) A risk assessment undertaken in accordance with Annex III; and
(b) A decision made within a predictable timeframe, not exceeding two hundred and seventy days.

7. Failure by a Party to communicate its decision according to paragraph 6 above, shall not imply its consent or refusal to the import of a living modified organism intended for direct use as food or feed, or for processing, unless otherwise specified by the Party.

8. Lack of scientific certainty due to insufficient relevant scientific information and knowledge regarding the extent of the potential adverse effects of a living modified organism on the conservation and sustainable use of biological diversity in the Party of import, taking also into account risks to human health, shall not prevent that Party from taking a decision, as appropriate, with regard to the import of that living modified organism intended for direct use as food or feed, or for processing, in order to avoid or minimize such potential adverse effects.

9. A Party may indicate its needs for financial and technical assistance and capacity-building with respect to living modified organisms intended for direct use as food or feed, or for processing. Parties shall cooperate to meet these needs in accordance with Articles 22 and 28.

Article 12
Review of Decisions

1. A Party of import may, at any time, in light of new scientific information on potential adverse effects on the conservation and sustainable use of biological diversity, taking also into account the risks to human health, review and change a decision regarding an

intentional transboundary movement. In such case, the Party shall, within thirty days, inform any notifier that has previously notified movements of the living modified organism referred to in such decision, as well as the Biosafety Clearing-House, and shall set out the reasons for its decision.

2. A Party of export or a notifier may request the Party of import to review a decision it has made in respect of it under Article 10 where the Party of export or the notifier considers that:

(a) A change in circumstances has occurred that may influence the outcome of the risk assessment upon which the decision was based; or
(b) Additional relevant scientific or technical information has become available.

3. The Party of import shall respond in writing to such a request within ninety days and set out the reasons for its decision.

4. The Party of import may, at its discretion, require a risk assessment for subsequent imports.

Article 13
Simplified Procedure

1. A Party of import may, provided that adequate measures are applied to ensure the safe intentional transboundary movement of living modified organisms in accordance with the objective of this Protocol, specify in advance to the Biosafety Clearing-House:

(a) Cases in which intentional transboundary movement to it may take place at the same time as the movement is notified to the Party of import; and
(b) Imports of living modified organisms to it to be exempted from the advance informed agreement procedure. Notifi-

cations under subparagraph (a) above, may apply to subsequent similar movements to the same Party.

2. The information relating to an intentional transboundary movement that is to be provided in the notifications referred to in paragraph 1 (a) above, shall be the information specified in Annex I.

Article 14
Bilateral, Regional and Multilateral agreements and Arrangements

1. Parties may enter into bilateral, regional and multilateral agreements and arrangements regarding intentional transboundary movements of living modified organisms, consistent with the objective of this Protocol and provided that such agreements and arrangements do not result in a lower level of protection than that provided for by the Protocol.

2. The Parties shall inform each other, through the Biosafety Clearing-House, of any such bilateral, regional and multilateral agreements and arrangements that they have entered into before or after the date of entry into force of this Protocol.

3. The provisions of this Protocol shall not affect intentional transboundary movements that take place pursuant to such agreements and arrangements as between the parties to those agreements or arrangements.

4. Any Party may determine that its domestic regulations shall apply with respect to specific imports to it and shall notify the Biosafety Clearing-House of its decision.

Article 15
Risk Assessment

1. Risk assessments undertaken pursuant to this Protocol shall be

carried out in a scientifically sound manner, in accordance with Annex III and taking into account recognized risk assessment techniques. Such risk assessments shall be based, at a minimum, on information provided in accordance with Article 8 and other available scientific evidence in order to identify and evaluate the possible adverse effects of living modified organisms on the conservation and sustainable use of biological diversity, taking also into account risks to human health.

2. The Party of import shall ensure that risk assessments are carried out for decisions taken under Article 10. It may require the exporter to carry out the risk assessment.

3. The cost of risk assessment shall be borne by the notifier if the Party of import so requires.

Article 16
Risk Management

1. The Parties shall, taking into account Article 8 (g) of the Convention, establish and maintain appropriate mechanisms, measures and strategies to regulate, manage and control risks identified in the risk assessment provisions of this Protocol associated with the use, handling and transboundary movement of living modified organisms.

2. Measures based on risk assessment shall be imposed to the extent necessary to prevent adverse effects of the living modified organism on the conservation and sustainable use of biological diversity, taking also into account risks to human health, within the territory of the Party of import.

3. Each Party shall take appropriate measures to prevent unintentional transboundary movements of living modified organisms, including such measures as requiring a risk assessment to be carried out prior to the first release of a living modified

organism.

4. Without prejudice to paragraph 2 above, each Party shall endeavour to ensure that any living modified organism, whether imported or locally developed, has undergone an appropriate period of observation that is commensurate with its life-cycle or generation time before it is put to its intended use.

5. Parties shall cooperate with a view to:

(a) Identifying living modified organisms or specific traits of living modified organisms that may have adverse effects on the conservation and sustainable use of biological diversity, taking also into account risks to human health; and
(b) Taking appropriate measures regarding the treatment of such living modified organisms or specific traits.

Article 17
Unintentional Transboundary Movements and Emergency Measures

1. Each Party shall take appropriate measures to notify affected or potentially affected States, the Biosafety Clearing-House and, where appropriate, relevant international organizations, when it knows of an occurrence under its jurisdiction resulting in a release that leads, or may lead, to an unintentional transboundary movement of a living modified organism that is likely to have significant adverse effects on the conservation and sustainable use of biological diversity, taking also into account risks to human health in such States. The notification shall be provided as soon as the Party knows of the above situation.

2. Each Party shall, no later than the date of entry into force of this Protocol for it, make available to the Biosafety Clearing-House the relevant details setting out its point of contact for the purposes of receiving notifications under this Article.

3. Any notification arising from paragraph 1 above, should include:

(a) Available relevant information on the estimated quantities and relevant characteristics and/or traits of the living modified organism;
(b) Information on the circumstances and estimated date of the release, and on the use of the living modified organism in the originating Party;
(c) Any available information about the possible adverse effects on the conservation and sustainable use of biological diversity, taking also into account risks to human health, as well as available information about possible risk management measures;
(d) Any other relevant information; and
(e) A point of contact for further information.

4. In order to minimize any significant adverse effects on the conservation and sustainable use of biological diversity, taking also into account risks to human health, each Party, under whose jurisdiction the release of the living modified organism referred to in paragraph 1 above, occurs, shall immediately consult the affected or potentially affected States to enable them to determine appropriate responses and initiate necessary action, including emergency measures.

Article 18
Handling, Transport, Packaging and Identification

1. In order to avoid adverse effects on the conservation and sustainable use of biological diversity, taking also into account risks to human health, each Party shall take necessary measures to require that living modified organisms that are subject to intentional transboundary movement within the scope of this Protocol are handled, packaged and transported under conditions

of safety, taking into consideration relevant international rules and standards.

2. Each Party shall take measures to require that documentation accompanying:

(a) Living modified organisms that are intended for direct use as food or feed, or for processing, clearly identifies that they "may contain" living modified organisms and are not intended for intentional introduction into the environment, as well as a contact point for further information. The Conference of the Parties serving as the meeting of the Parties to this Protocol shall take a decision on the detailed requirements for this purpose, including specification of their identity and any unique identification, no later than two years after the date of entry into force of this Protocol;

(b) Living modified organisms that are destined for contained use clearly identifies them as living modified organisms; and specifies any requirements for the safe handling, storage, transport and use, the contact point for further information, including the name and address of the individual and institution to whom the living modified organisms are consigned; and

(c) Living modified organisms that are intended for intentional introduction into the environment of the Party of import and any other living modified organisms within the scope of the Protocol, clearly identifies them as living modified organisms; specifies the identity and relevant traits and/or characteristics, any requirements for the safe handling, storage, transport and use, the contact point for further information and, as appropriate, the name and address of the importer and exporter; and contains a declaration that the movement is in conformity with the requirements of this Protocol applicable to the exporter.

3. The Conference of the Parties serving as the meeting of the

Parties to this Protocol shall consider the need for and modalities of developing standards with regard to identification, handling, packaging and transport practices, in consultation with other relevant international bodies.

Article 19
Competent National Authorities and National Focal Points

1. Each Party shall designate one national focal point to be responsible on its behalf for liaison with the Secretariat. Each Party shall also designate one or more competent national authorities, which shall be responsible for performing the administrative functions required by this Protocol and which shall be authorized to act on its behalf with respect to those functions. A Party may designate a single entity to fulfil the functions of both focal point and competent national authority.

2. Each Party shall, no later than the date of entry into force of this Protocol for it, notify the Secretariat of the names and addresses of its focal point and its competent national authority or authorities. Where a Party designates more than one competent national authority, it shall convey to the Secretariat, with its notification thereof, relevant information on the respective responsibilities of those authorities. Where applicable, such information shall, at a minimum, specify which competent authority is responsible for which type of living modified organism. Each Party shall forthwith notify the Secretariat of any changes in the designation of its national focal point or in the name and address or responsibilities of its competent national authority or authorities.

3. The Secretariat shall forthwith inform the Parties of the notifications it receives under paragraph 2 above, and shall also make such information available through the Biosafety Clearing-House.

Article 20
Information Sharing and the Biosafety Clearing-House

1. A Biosafety Clearing-House is hereby established as part of the Clearing-House. mechanism under Article 18, paragraph 3, of the Convention, in order to:

(a) Facilitate the exchange of scientific, technical, environmental and legal information on, and experience with, living modified organisms; and
(b) Assist Parties to implement the Protocol, taking into account the special needs of developing country Parties, in particular the least developed and small island developing States among them, and countries with economies in transition as well as countries that are centres of ori gin and centres of genetic diversity.

2. The Biosafety Clearing-House shall serve as a means through which information is made available for the purposes of paragraph 1 above. It shall provide access to information made available by the Parties relevant to the implementation of the Protocol. It shall also provide access, where possible, to other international biosafety information exchange mechanisms.

3. Without prejudice to the protection of confidential information, each Party shall make available to the Biosafety Clearing-House any information required to be made available to the Biosafety Clearing-House under this Protocol, and:

(a) Any existing laws, regulations and guidelines for implementation of the Protocol, as well as information required by the Parties for the advance informed agreement procedure;
(b) Any bilateral, regional and multilateral agreements and arrangements;
(c) Summaries of its risk assessments or environmental reviews

of living modified organisms generated by its regulatory process, and carried out in accordance with Article 15, including, where appropriate, relevant information regarding products thereof, namely, processed materials that are of living modified organism origin, containing detectable novel combinations of replicable genetic material obtained through the use of modern biotechnology;

(d) Its final decisions regarding the importation or release of living modified organisms; and

(e) Reports submitted by it pursuant to Article 33, including those on implementation of the advance informed agreement procedure.

4. The modalities of the operation of the Biosafety Clearing-House, including reports on its activities, shall be considered and decided upon by the Conference of the Parties serving as the meeting of the Parties to this Protocol at its first meeting, and kept under review thereafter.

Article 21
Confidentional Information

1. The Party of import shall permit the notifier to identify information submitted under the procedures of this Protocol or required by the Party of import as part of the advance informed agreement procedure of the Protocol that is to be treated as confidential. Justification shall be given in such cases upon request.

2. The Party of import shall consult the notifier if it decides that information identified by the notifier as confidential does not qualify for such treatment and shall, prior to any disclosure, inform the notifier of its decision, providing reasons on request, as well as an opportunity for consultation and for an internal review of the decision prior to disclosure.

3. Each Party shall protect confidential information received under this Protocol, including any confidential information received in the context of the advance informed agreement procedure of the Protocol. Each Party shall ensure that it has procedures to protect such information and shall protect the confidentiality of such information in a manner no less favourable than its treatment of confidential information in connection with domestically produced living modified organisms.

4. The Party of import shall not use such information for a commercial purpose, except with the written consent of the notifier.

5. If a notifier withdraws or has withdrawn a notification, the Party of import shall respect the confidentiality of commercial and industrial information, including research and development information as well as information on which the Party and the notifier disagree as to its confidentiality.

6. Without prejudice to paragraph 5 above, the following information shall not be considered confidential:

(a) The name and address of the notifier;
(b) A general description of the living modified organism or organisms;
(c) A summary of the risk assessment of the effects on the conservation and sustainable use of biological diversity, taking also into account risks to human health; and
(d) Any methods and plans for emergency response.

Article 22
Capacity-Building

1. The Parties shall cooperate in the development and/or strengthening of human resources and institutional capacities in biosafety, including biotechnology to the extent that it is required

for biosafety, for the purpose of the effective implementation of this Protocol, in developing country Parties, in particular the least developed and small island developing States among them, and in Parties with economies in transition, including through existing global, regional, subregional and national institutions and organizations and, as appropriate, through facilitating private sector involvement.

2. For the purposes of implementing paragraph 1 above, in relation to cooperation, the needs of developing country Parties, in particular the least developed and small island developing States among them, for financial resources and access to and transfer of technology and know-how in accordance with the relevant provisions of the Convention, shall be taken fully into account for capacity-building in biosafety. Cooperation in capacitybuilding shall, subject to the different situation, capabilities and requirements of each Party, include scientific and technical training in the proper and safe management of biotechnology, and in the use of risk assessment and risk management for biosafety, and the enhancement of technological and institutional capacities in biosafety. The needs of Parties with economies in transition shall also be taken fully into account for such capacity-building in biosafety.

Article 23
Public Awareness and Participation

1. The Parties shall:

(a) Promote and facilitate public awareness, education and participation concerning the safe transfer, handling and use of living modified organisms in relation to the conservation and sustainable use of biological diversity, taking also into account risks to human health. In doing so, the Parties shall cooperate, as appropriate, with other States and international bodies;

(b) Endeavour to ensure that public awareness and education encompass access to information on living modified organisms identified in accordance with this Protocol that may be imported.

2. The Parties shall, in accordance with their respective laws and regulations, consult the public in the decision-making process regarding living modified organisms and shall make the results of such decisions available to the public, while respecting confidential information in accordance with Article 21.

3. Each Party shall endeavour to inform its public about the means of public access to the Biosafety Clearing-House.

Article 24
Non-Parties

1. Transboundary movements of living modified organisms between Parties and non-Parties shall be consistent with the objective of this Protocol. The Parties may enter into bilateral, regional and multilateral agreements and arrangements with non-Parties regarding such transboundary movements.

2. The Parties shall encourage non-Parties to adhere to this Protocol and to contribute appropriate information to the Biosafety Clearing-House on living modified organisms released in, or moved into or out of, areas within their national jurisdictions.

Article 25
Illegal Transboundary Movements

1. Each Party shall adopt appropriate domestic measures aimed at preventing and, if appropriate, penalizing transboundary movements of living modified organisms carried out in contravention of

its domestic measures to implement this Protocol. Such movements shall be deemed illegal transboundary movements.

2. In the case of an illegal transboundary movement, the affected Party may request the Party of origin to dispose, at its own expense, of the living modified organism in question by repatriation or destruction, as appropriate.

3. Each Party shall make available to the Biosafety Clearing-House information concerning cases of illegal transboundary movements pertaining to it.

Article 26
Socio-Economic Considerations

1. The Parties, in reaching a decision on import under this Protocol or under its domestic measures implementing the Protocol, may take into account, consistent with their international obligations, socio-economic considerations arising from the impact of living modified organisms on the conservation and sustainable use of biological diversity, especially with regard to the value of biological diversity to indigenous and local communities.

2. The Parties are encouraged to cooperate on research and information exchange on any socio-economic impacts of living modified organisms, especially on indigenous and local communities.

Article 27
Liability and Redress

The Conference of the Parties serving as the meeting of the Parties to this Protocol shall, at its first meeting, adopt a process with respect to the appropriate elaboration of international rules and procedures in the field of liability and redress for damage resulting from transboundary movements of living modified organisms, analysing and taking due account of the ongoing processes in international law on these matters, and shall endeavour to complete this process within four years.

Article 28
Financial Mechanism and Resources

1. In considering financial resources for the implementation of this Protocol, the Parties shall take into account the provisions of Article 20 of the Convention.

2. The financial mechanism established in Article 21 of the Convention hall, through the institutional structure entrusted with its operation, be the financial mechanism for this Protocol.

3. Regarding the capacity-building referred to in Article 22 of this Protocol, the Conference of the Parties serving as the meeting of the Parties to this Protocol, in providing guidance with respect to the financial mechanism referred to in paragraph 2 above, for consideration by the Conference of the Parties, shall take into account the need for financial resources by developing country Parties, in particular the least developed and the small island developing States among them.

4. In the context of paragraph 1 above, the Parties shall also take into account the needs of the developing country Parties, in particular the least developed and the small island developing States among them, and of the Parties with economies in transition, in their efforts to identify and implement their capacity -building requirements for the purposes of the implementation of this Protocol.

5. The guidance to the financial mechanism of the Convention in relevant decisions of the Conference of the Parties, including those agreed before the adoption of this Protocol, shall apply, mutatis

mutandis, to the provisions of this Article.

6. The developed country Parties may also provide, and the developing country Parties and the Parties with economies in transition avail themselves of, financial and technological resources for the implementation of the provisions of this Protocol through bilateral, regional and multilateral channels.

Article 29
Confidence of the Parties Serving as the Meeting of the Parties to this Protocol

1. The Conference of the Parties shall serve as the meeting of the Parties to this Protocol.

2. Parties to the Convention that are not Parties to this Protocol may participate as observers in the proceedings of any meeting of the Conference of the Parties serving as the meeting of the Parties to this Protocol. When the Conference of the Parties serves as the meeting of the Parties to this Protocol, decisions under this Protocol shall be taken only by those that are Parties to it.

3. When the Conference of the Parties serves as the meeting of the Parties to this Protocol, any member of the bureau of the Conference of the Parties representing a Party to the Convention but, at that time, not a Party to this Protocol, shall be substituted by a member to be elected by and from among the Parties to this Protocol.

4. The Conference of the Parties serving as the meeting of the Parties to this Protocol shall keep under regular review the implementation of this Protocol and shall make, within its mandate, the decisions necessary to promote its effective implementation. It shall perform the functions assigned to it by this Protocol and shall:

(a) Make recommendations on any matters necessary for the implementation of this Protocol;
(b) Establish such subsidiary bodies as are deemed necessary for the implementation of this Protocol;
(c) Seek and utilize, where appropriate, the services and cooperation of, and information provided by, competent international organizations and intergovernmental and non-governmental bodies;
(d) Establish the form and the intervals for transmitting the information to be submitted in accordance with Article 33 of this Protocol and consider such information as well as reports submitted by any subsidiary body;
(e) Consider and adopt, as required, amendments to this Protocol and its annexes, as well as any additional annexes to this Protocol, that are deemed necessary for the implementation of this Protocol; and
(f) Exercise such other functions as may be required for the implementation of this Protocol.

5. The rules of procedure of the Conference of the Parties and financial rules of the Convention shall be applied, mutatis mutandis, under this Protocol, except as may be otherwise decided by consensus by the Conference of the Parties serving as the meeting of the Parties to this Protocol.

6. The first meeting of the Conference of the Parties serving as the meeting of the Parties to this Protocol shall be convened by the Secretariat in conjunction with the first meeting of the Conference of the Parties that is scheduled after the date of the entry into force of this Protocol. Subsequent ordinary meetings of the Conference of the Parties serving as the meeting of the Parties to this Protocol shall be held in conjunction with ordinary meetings of the Conference of the Parties, unless otherwise decided by the Conference of the Parties serving as the meeting of the Parties to this Protocol.

7. Extraordinary meetings of the Conference of the Parties serving as the meeting of the Parties to this Protocol shall be held at such other times as may be deemed necessary by the Conference of the Parties serving as the meeting of the Parties to this Protocol, or at the written request of any Party, provided that, within six months of the request being communicated to the Parties by the Secretariat, it is supported by at least one third of the Parties.

8. The United Nations, its specialized agencies and the International Atomic Energy Agency, as well as any State member thereof or observers thereto not party to the Convention, may be represented as observers at meetings of the Conference of the Parties serving as the meeting of the Parties to this Protocol. Any body or agency, whether national or international, governmental or nongovernmental, that is qualified in matters covered by this Protocol and that has informed the Secretariat of its wish to be represented at a meeting of the Conference of the Parties serving as a meeting of the Parties to this Protocol as an observer, may be so admitted, unless at least one third of the Parties present object. Except as otherwise provided in this Article, the admission and participation of observers shall be subject to the rules of procedure, as referred to in paragraph 5 above.

Article 30
Subsidiary Bodies

1. Any subsidiary body established by or under the Convention may, upon a decision by the Conference of the Parties serving as the meeting of the Parties to this Protocol, serve the Protocol, in which case the meeting of the Parties shall specify which functions that body shall exercise.

2. Parties to the Convention that are not Parties to this Protocol may participate as observers in the proceedings of any meeting of

any such subsidiary bodies. When a subsidiary body of the Convention serves as a subsidiary body to this Protocol, decisions under the Protocol shall be taken only by the Parties to the Protocol.

3. When a subsidiary body of the Convention exercises its functions with regard to matters concerning this Protocol, any member of the bureau of that subsidiary body representing a Party to the Convention but, at that time, not a Party to the Protocol, shall be substituted by a member to be elected by and from among the Parties to the Protocol.

Article 31
Secretariat

1. The Secretariat established by Article 24 of the Convention shall serve as the secretariat to this Protocol.

2. Article 24, paragraph 1, of the Convention on the functions of the Secretariat shall apply, mutatis mutandis, to this Protocol.

3. To the extent that they are distinct, the costs of the secretariat services for this Protocol shall be met by the Parties hereto. The Conference of the Parties serving as the meeting of the Parties to this Protocol shall, at its first meeting, decide on the necessary budgetary arrangements to this end.

Article 32
Relationship with the Convention

Except as otherwise provided in this Protocol, the provisions of the Convention relating to its protocols shall apply to this Protocol.

Article 33
Monitoring and Reporting

Each Party shall monitor the implementation of its obligations under this Protocol, and shall, at intervals to be determined by the Conference of the Parties serving as the meeting of the Parties to this Protocol, report to the Conference of the Parties serving as the meeting of the Parties to this Protocol on measures that it has taken to implement the Protocol.

Article 34
Compliance

The Conference of the Parties serving as the meeting of the Parties to this Protocol shall, at its first meeting, consider and approve cooperative procedures and institutional mechanisms to promote compliance with the provisions of this Protocol and to address cases of non-compliance. These procedures and mechanisms shall include provisions to offer advice or assistance, where appropriate. They shall be separate from, and without prejudice to, the dispute settlement procedures and mechanisms established by Article 27 of the Convention.

Article 35
Assessment and Review

The Conference of the Parties serving as the meeting of the Parties to this Protocol shall undertake, five years after the entry into force of this Protocol and at least every five years thereafter, an evaluation of the effectiveness of the Protocol, including an assessment of its procedures and annexes.

Article 36
Signature

This Protocol shall be open for signature at the United Nations Office at Nairobi by States and regional economic integration organizations from 15 to 26 May 2000, and at United Nations Headquarters in New York from 5 June 2000 to 4 June 2001.

Article 37
Entry into Force

1. This Protocol shall enter into force on the ninetieth day after the date of deposit of the fiftieth instrument of ratification, acceptance, approval or accession by States or regional economic integration organizations that are Parties to the Convention.

2. This Protocol shall enter into force for a State or regional economic integration organization that ratifies, accepts or approves this Protocol or accedes thereto after its entry into force pursuant to paragraph 1 above, on the ninetieth day after the date on which that State or regional economic integration organization deposits its instrument of ratification, acceptance, approval or accession, or on the date on which the Convention enters into force for that State or regional economic integration organization, whichever shall be the later.

3. For the purposes of paragraphs 1 and 2 above, any instrument deposited by a regional economic integration organization shall not be counted as additional to those deposited by member States of such organization.

Article 38
Reservations

No reservations may be made to this Protocol.

Article 39
Withdrawal

1. At any time after two years from the date on which this Protocol has entered into force for a Party, that Party may withdraw from the Protocol by giving written notification to the Depositary.

2. Any such withdrawal shall take place upon expiry of one year after the date of its receipt by the Depositary, or on such later date as may be specified in the notification of the withdrawal.

Article 40
Authentic Texts

The original of this Protocol, of which the Arabic, Chinese, English, French, Russian and Spanish texts are equally authentic, shall be deposited with the Secretary-General of the United Nations.

IN WITNESS WHEREOF the undersigned, being duly authorized to that effect, have signed this Protocol.

DONE at Montreal on this twenty-ninth day of January, two thousand.

Annex I
Information Required in Notifications under Articles 8, 10 and 13

(a) Name, address and contact details of the exporter.
(b) Name, address and contact details of the importer.
(c) Name and identity of the living modified organism, as well as the domestic classification, if any, of the biosafety level of the living modified organism in the State of export.

(d) Intended date or dates of the transboundary movement, if known.
(e) Taxonomic status, common name, point of collection or acquisition, and characteristics of recipient organism or parental organisms related to biosafety.
(f) Centres of origin and centres of genetic diversity, if known, of the recipient organism and/or the parental organisms and a description of the habitats where the organisms may persist or proliferate.
(g) Taxonomic status, common name, point of collection or acquisition, and characteristics of the donor organism or organisms related to biosafety.
(h) Description of the nucleic acid or the modification introduced, the technique used, and the resulting characteristics of the living modified organism.
(i) Intended use of the living modified organism or products thereof, namely, processed materials that are of living modified organism origin, containing detectable novel combinations of replicable genetic material obtained through the use of modern biotechnology.
(j) Quantity or volume of the living modified organism to be transferred.
(k) A previous and existing risk assessment report consistent with Annex III.
(l) Suggested methods for the safe handling, storage, transport and use, including packaging, labelling, documentation, disposal and contingency procedures, where appropriate.
(m) Regulatory status of the living modified organism within the State of export (for example, whether it is prohibited in the State of export, whether there are other restrictions, or whether it has been approved for general release) and, if the living modified organism is banned in the State of export, the reason or reasons for the ban.
(n) Result and purpose of any notification by the exporter to

other States regarding the living modified organism to be transferred.

(o) A declaration that the above-mentioned information is factually correct.

Annex II

Information Required Concerning Living Modified Organisms Intended For Direct Use as Food or Feed, or For Processing under Article 11

(a) The name and contact details of the applicant for a decision for domestic use.

(b) The name and contact details of the authority responsible for the decision.

(c) Name and identity of the living modified organism.

(d) Description of the gene modification, the technique used, and the resulting characteristics of the living modified organism.

(e) Any unique identification of the living modified organism.

(f) Taxonomic status, common name, point of collection or acquisition, and characteristics of recipient organism or parental organisms related to biosafety.

(g) Centres of origin and centres of genetic diversity, if known, of the recipient organism and/or the parental organisms and a description of the habitats where the organisms may persist or proliferate.

(h) Taxonomic status, common name, point of collection or acquisition, and characteristics of the donor organism or organisms related to biosafety.

(i) Approved uses of the living modified organism.

(j) A risk assessment report consistent with Annex III.

(k) Suggested methods for the safe handling, storage, transport and use, including packaging, labelling, documentation, disposal and contingency procedures, where appropriate.

Annex III
Risk Assessment

Objective

1. The objective of risk assessment, under this Protocol, is to identify and evaluate the potential adverse effects of living modified organisms on the conservation and sustainable use of biological diversity in the likely potential receiving environment, taking also into account risks to human health.

Use of risk assessment

2. Risk assessment is, inter alia, used by competent authorities to make informed decisions regarding living modified organisms.

General principles

3. Risk assessment should be carried out in a scientifically sound and transparent manner, and can take into account expert advice of, and guidelines developed by, relevant international organizations.

4. Lack of scientific knowledge or scientific consensus should not necessarily be interpreted as indicating a particular level of risk, an absence of risk, or an acceptable risk.

5. Risks associated with living modified organisms or products thereof, namely, processed materials that are of living modified organism origin, containing detectable novel combinations of replicable genetic material obtained through the use of modern biotechnology, should be considered in the context of the risks posed by the non-modified recipients or parental organisms in the likely potential receiving environment.

6. Risk assessment should be carried out on a case-by-case basis. The required information may vary in nature and level of detail from case to case, depending on the living modified organism concerned, its intended use and the likely potential receiving environment.

Methodology

7. The process of risk assessment may on the one hand give rise to a need for further information about specific subjects, which may be identified and requested during the assessment process, while on the other hand information on other subjects may not be relevant in some instances.

8. To fulfil its objective, risk assessment entails, as appropriate, the following steps:

(a) An identification of any novel genotypic and phenotypic characteristics associated with the living modified organism that may have adverse effects on biological diversity in the likely potential receiving environment, taking also into account risks to human health;

(b) An evaluation of the likelihood of these adverse effects being realized, taking into account the level and kind of exposure of the likely potential receiving environment to the living modified organism;

(c) An evaluation of the consequences should these adverse effects be realized;

(d) An estimation of the overall risk posed by the living modified organism based on the evaluation of the likelihood and consequences of the identified adverse effects being realized;

(e) A recommendation as to whether or not the risks are acceptable or manageable, including, where necessary, identification of strategies to manage these risks; and

(f) Where there is uncertainty regarding the level of risk, it may be addressed by requesting further information on the specific issues of concern or by implementing appropriate risk management strategies and/or monitoring the living modified organism in the receiving environment.

Points to consider

9. Depending on the case, risk assessment takes into account the relevant technical and scientific details regarding the characteristics of the following subjects:

(a) Recipient organism or parental organisms. The biological characteristics of the recipient organism or parental organisms, including information on taxonomic status, common name, origin, centres of origin and centres of genetic diversity, if known, and a description of the habitat where the organisms may persist or proliferate;
(b) Donor organism or organisms. Taxonomic status and common name, source, and the relevant biological characteristics of the donor organisms; source or origin, and its host range;
(d) Insert or inserts and/or characteristics of modification. Genetic characteristics of the inserted nucleic acid and the function it specifies, and/or characteristics of the modification introduced;
(e) Living modified organism. Identity of the living modified organism, and the differences between the biological characteristics of the living modified organism and those of the recipient organism or parental organisms;
(f) Detection and identification of the living modified organism. Suggested detection and identification methods and their specificity, sensitivity and reliability;
(g) Information relating to the intended use. Information

relating to the intended use of the living modified organism, including new or changed use compared to the recipient organism or parental organisms; and

(h) Receiving environment. Information on the location, geographical, climatic and ecological characteristics, including relevant information on biological diversity and centres of origin of the likely potential receiving environment.

부 록

우리나라 저탄소녹색성장기본법 (2010)

저탄소 녹색성장 기본법

[시행 2010. 4. 14] [법률 제9931호, 2010. 1. 13. 제정]

제1장 총 칙

제1조 【목적】 이 법은 경제와 환경의 조화로운 발전을 위하여 저탄소(低炭素) 녹색성장에 필요한 기반을 조성하고 녹색기술과 녹색산업을 새로운 성장동력으로 활용함으로써 국민경제의 발전을 도모하며 저탄소 사회 구현을 통하여 국민의 삶의 질을 높이고 국제사회에서 책임을 다하는 성숙한 선진 일류국가로 도약하는 데 이바지함을 목적으로 한다.

제2조 【정의】 이 법에서 사용하는 용어의 뜻은 다음과 같다.

1. "저탄소"란 화석연료(化石燃料)에 대한 의존도를 낮추고 청정에너지의 사용 및 보급을 확대하며 녹색기술 연구개발, 탄소흡수원 확충 등을 통하여 온실가스를 적정수준 이하로 줄이는 것을 말한다.
2. "녹색성장"이란 에너지와 자원을 절약하고 효율적으로 사용하여 기후변화와 환경훼손을 줄이고 청정에너지와 녹색기술의 연구개발을 통하여 새로운 성장동력을 확보하며 새로운 일자리를 창출해 나가는 등 경제와 환경이 조화를 이루는 성장을 말한다.
3. "녹색기술"이란 온실가스 감축기술, 에너지 이용 효율화 기술, 청정생산기술, 청정에너지 기술, 자원순환 및 친환경 기술(관련 융합기술을 포함한다) 등 사회·경제 활동의

전 과정에 걸쳐 에너지와 자원을 절약하고 효율적으로 사용하여 온실가스 및 오염물질의 배출을 최소화하는 기술을 말한다.

4. "녹색산업"이란 경제·금융·건설·교통물류·농림수산·관광 등 경제활동 전반에 걸쳐 에너지와 자원의 효율을 높이고 환경을 개선할 수 있는 재화(財貨)의 생산 및 서비스의 제공 등을 통하여 저탄소 녹색성장을 이루기 위한 모든 산업을 말한다.
5. "녹색제품"이란 에너지·자원의 투입과 온실가스 및 오염물질의 발생을 최소화하는 제품을 말한다.
6. "녹색생활"이란 기후변화의 심각성을 인식하고 일상생활에서 에너지를 절약하여 온실가스와 오염물질의 발생을 최소화하는 생활을 말한다.
7. "녹색경영"이란 기업이 경영활동에서 자원과 에너지를 절약하고 효율적으로 이용하며 온실가스 배출 및 환경오염의 발생을 최소화하면서 사회적, 윤리적 책임을 다하는 경영을 말한다.
8. "지속가능발전"이란 「지속가능발전법」 제2조 제2호에 따른 지속가능발전을 말한다.
9. "온실가스"란 이산화탄소(CO2), 메탄(CH4), 아산화질소(N2O), 수소불화탄소(HFCs), 과불화탄소(PFCs), 육불화황(SF6) 및 그 밖에 대통령령으로 정하는 것으로 적외선 복사열을 흡수하거나 재방출하여 온실효과를 유발하는 대기 중의 가스 상태의 물질을 말한다.
10. "온실가스 배출"이란 사람의 활동에 수반하여 발생하는 온실가스를 대기 중에 배출·방출 또는 누출시키는 직접배출과 다른 사람으로부터 공급된 전기 또는 열(연료 또는 전기를 열원으로 하는 것만 해당한다)을 사용함으로써 온실가스가 배출되도록 하는 간접배출을 말한다.

11. "지구온난화"란 사람의 활동에 수반하여 발생하는 온실가스가 대기 중에 축적되어 온실가스 농도를 증가시킴으로써 지구 전체적으로 지표 및 대기의 온도가 추가적으로 상승하는 현상을 말한다.
12. "기후변화"란 사람의 활동으로 인하여 온실가스의 농도가 변함으로써 상당 기간 관찰되어 온 자연적인 기후변동에 추가적으로 일어나는 기후체계의 변화를 말한다.
13. "자원순환"이란 「자원의 절약과 재활용촉진에 관한 법률」 제2조 제1호에 따른 자원순환을 말한다.
14. "신·재생에너지"란 「신에너지 및 재생에너지 개발·이용·보급 촉진법」 제2조 제1호에 따른 신에너지 및 재생에너지를 말한다.
15. "에너지 자립도"란 국내 총소비에너지량에 대하여 신·재생에너지 등 국내 생산에너지량 및 우리나라가 국외에서 개발(지분 취득을 포함한다)한 에너지량을 합한 양이 차지하는 비율을 말한다.

제3조【저탄소 녹색성장 추진의 기본원칙】 저탄소 녹색성장은 다음 각 호의 기본원칙에 따라 추진되어야 한다.

1. 정부는 기후변화·에너지·자원 문제의 해결, 성장동력 확충, 기업의 경쟁력 강화, 국토의 효율적 활용 및 쾌적한 환경 조성 등을 포함하는 종합적인 국가 발전전략을 추진한다.
2. 정부는 시장기능을 최대한 활성화하여 민간이 주도하는 저탄소 녹색성장을 추진한다.
3. 정부는 녹색기술과 녹색산업을 경제성장의 핵심 동력으로 삼고 새로운 일자리를 창출·확대할 수 있는 새로운 경제체제를 구축한다.
4. 정부는 국가의 자원을 효율적으로 사용하기 위하여 성장

잠재력과 경쟁력이 높은 녹색기술 및 녹색산업 분야에 대한 중점 투자 및 지원을 강화한다.

5. 정부는 사회·경제 활동에서 에너지와 자원 이용의 효율성을 높이고 자원순환을 촉진한다.
6. 정부는 자연자원과 환경의 가치를 보존하면서 국토와 도시, 건물과 교통, 도로·항만·상하수도 등 기반시설을 저탄소 녹색성장에 적합하게 개편한다.
7. 정부는 환경오염이나 온실가스 배출로 인한 경제적 비용이 재화 또는 서비스의 시장가격에 합리적으로 반영되도록 조세(租稅)체계와 금융체계를 개편하여 자원을 효율적으로 배분하고 국민의 소비 및 생활 방식이 저탄소 녹색성장에 기여하도록 적극 유도한다. 이 경우 국내산업의 국제경쟁력이 약화되지 않도록 고려하여야 한다.
8. 정부는 국민 모두가 참여하고 국가기관, 지방자치단체, 기업, 경제단체 및 시민단체가 협력하여 저탄소 녹색성장을 구현하도록 노력한다.
9. 정부는 저탄소 녹색성장에 관한 새로운 국제적 동향(動向)을 조기에 파악·분석하여 국가 정책에 합리적으로 반영하고, 국제사회의 구성원으로서 책임과 역할을 성실히 이행하여 국가의 위상과 품격을 높인다.

제4조【국가의 책무】 ① 국가는 정치·경제·사회·교육·문화 등 국정의 모든 부문에서 저탄소 녹색성장의 기본원칙이 반영될 수 있도록 노력하여야 한다.

② 국가는 각종 정책을 수립할 때 경제와 환경의 조화로운 발전 및 기후변화에 미치는 영향 등을 종합적으로 고려하여야 한다.

③ 국가는 지방자치단체의 저탄소 녹색성장 시책을 장려하고 지원하며, 녹색성장의 정착·확산을 위하여 사업자와 국민, 민간단체에 정보의 제공 및 재정 지원 등 필요한 조치를 할 수 있다.

④ 국가는 에너지와 자원의 위기 및 기후변화 문제에 대한 대응책을 정기적으로 점검하여 성과를 평가하고 국제협상의 동향 및 주요 국가의 정책을 분석하여 적절한 대책을 마련하여야 한다.

⑤ 국가는 국제적인 기후변화대응 및 에너지·자원 개발협력에 능동적으로 참여하고, 개발도상국가에 대한 기술적·재정적 지원을 할 수 있다.

제5조【지방자치단체의 책무】 ① 지방자치단체는 저탄소 녹색성장 실현을 위한 국가시책에 적극 협력하여야 한다.

② 지방자치단체는 저탄소 녹색성장대책을 수립·시행할 때 해당 지방자치단체의 지역적 특성과 여건을 고려하여야 한다.

③ 지방자치단체는 관할구역 내에서의 각종 계획 수립과 사업의 집행과정에서 그 계획과 사업이 저탄소 녹색성장에 미치는 영향을 종합적으로 고려하고, 지역주민에게 저탄소 녹색성장에 대한 교육과 홍보를 강화하여야 한다.

④ 지방자치단체는 관할구역 내의 사업자, 주민 및 민간단체의 저탄소 녹색성장을 위한 활동을 장려하기 위하여 정보 제공, 재정 지원 등 필요한 조치를 강구하여야 한다.

제6조【사업자의 책무】 ① 사업자는 녹색경영을 선도하여야 하며 기업활동의 전 과정에서 온실가스와 오염물질의 배출을 줄이고 녹색기술 연구개발과 녹색산업에 대한 투자 및 고용을 확대하는 등 환경에 관한 사회적·윤리적 책임을 다하여야 한다.

② 사업자는 정부와 지방자치단체가 실시하는 저탄소 녹색성장에 관한 정책에 적극 참여하고 협력하여야 한다.

제7조【국민의 책무】 ① 국민은 가정과 학교 및 직장 등에서 녹색생활을 적극 실천하여야 한다.

② 국민은 기업의 녹색경영에 관심을 기울이고 녹색제품의 소비 및 서비스 이용을 증대함으로써 기업의 녹색경영을 촉진한다.

③ 국민은 스스로가 인류가 직면한 심각한 기후변화, 에너지·자원 위기의 최종적인 문제해결자임을 인식하여 건강하고 쾌적한 환경을 후손에게 물려주기 위하여 녹색생활 운동에 적극 참여하여야 한다.

제8조【다른 법률과의 관계】 ① 저탄소 녹색성장에 관하여는 다른 법률에 우선하여 이 법을 적용한다.

② 저탄소 녹색성장과 관련되는 다른 법률을 제정하거나 개정하는 경우에는 이 법의 목적과 기본원칙에 맞도록 하여야 한다.

③ 국가와 지방자치단체가 다른 법령에 따라 수립하는 행정계획과 정책은 제3조에 따른 저탄소 녹색성장 추진의 기본원칙 및 제9조에 따른 저탄소 녹색성장 국가전략과 조화를 이루도록 하여야 한다.

제 2 장 저탄소 녹색성장 국가전략

제9조【저탄소 녹색성장 국가전략】 ① 정부는 국가의 저탄소 녹색성장을 위한 정책목표·추진전략·중점추진과제 등을 포함하는 저탄소 녹색성장 국가전략(이하 "녹색성장국가전략"이라 한다)을 수립·시행하여야 한다.

② 녹색성장국가전략에는 다음 각 호의 사항이 포함되어야 한다.

1. 제22조에 따른 녹색경제 체제의 구현에 관한 사항
2. 녹색기술·녹색산업에 관한 사항
3. 기후변화대응 정책, 에너지 정책 및 지속가능발전 정책에 관한 사항
4. 녹색생활, 제51조에 따른 녹색국토, 제53조에 따른 저탄소 교통체계 등에 관한 사항

5. 기후변화 등 저탄소 녹색성장과 관련된 국제협상 및 국제협력에 관한 사항
6. 그 밖에 재원조달, 조세·금융, 인력양성, 교육·홍보 등 저탄소 녹색성장을 위하여 필요하다고 인정되는 사항

③ 정부는 녹색성장국가전략을 수립하거나 변경하려는 경우 제14조에 따른 녹색성장위원회의 심의 및 국무회의의 심의를 거쳐야 한다. 다만, 대통령령으로 정하는 경미한 사항을 변경하는 경우에는 그러하지 아니한다.

제10조【중앙행정기관의 추진계획 수립·시행】 ① 중앙행정기관의 장은 녹색성장국가전략을 효율적·체계적으로 이행하기 위하여 대통령령으로 정하는 바에 따라 소관 분야의 추진계획(이하 "중앙추진계획"이라 한다)을 수립·시행하여야 한다.

② 중앙행정기관의 장은 중앙추진계획을 수립하거나 변경하는 때에는 대통령령으로 정하는 바에 따라 제14조에 따른 녹색성장위원회에 보고하여야 한다. 다만, 대통령령으로 정하는 경미한 사항을 변경하는 경우에는 그러하지 아니하다.

제11조【지방자치단체의 추진계획 수립·시행】 ① 특별시장·광역시장·도지사 또는 특별자치도지사(이하 "시·도지사"라 한다)는 해당 지방자치단체의 저탄소 녹색성장을 촉진하기 위하여 대통령령으로 정하는 바에 따라 녹색성장국가전략과 조화를 이루는 지방녹색성장 추진계획(이하 "지방추진계획"이라 한다)을 수립·시행하여야 한다.

② 시·도지사는 지방추진계획을 수립하거나 변경하는 때에는 제20조에 따른 지방녹색성장위원회의 심의를 거친 후 지방의회에 보고하고 지체 없이 이를 제14조에 따른 녹색성장위원회에 제출하여야 한다. 다만, 대통령령으로 정하는 경미한 사항을 변경하는 경우에는 그러하지 아니하다.

제12조【추진상황 점검 및 평가】 ① 국무총리는 대통령령으로 정하는 바에 따라 녹색성장국가전략과 중앙추진계획의 이행사항을 점검·평가하여야 한다. 이 경우 국무총리는 평가의 절차, 기준, 결과 등에 대하여 제14조에 따른 녹색성장위원회와 협의하여야 한다.

② 시·도지사는 대통령령으로 정하는 바에 따라 지방추진계획의 이행상황을 점검·평가하여 그 결과를 지방의회에 보고하고 지체 없이 이를 제14조에 따른 녹색성장위원회에 제출하여야 한다.

제13조【정책에 관한 의견제시】 ① 제14조에 따른 녹색성장위원회는 제12조에 따른 추진상황 점검·평가 결과 등에 따라 필요하다고 인정되는 경우에는 관계 중앙행정기관의 장 또는 시·도지사에게 의견을 제시할 수 있다.

② 제1항에 따른 의견을 제시받은 관계 중앙행정기관의 장 또는 시·도지사는 해당 기관의 정책 등에 이를 반영하기 위하여 노력하여야 한다.

제 3 장 녹색성장위원회 등

제14조【녹색성장위원회의 구성 및 운영】 ① 국가의 저탄소 녹색성장과 관련된 주요 정책 및 계획과 그 이행에 관한 사항을 심의하기 위하여 대통령 소속으로 녹색성장위원회(이하 “위원회”라 한다)를 둔다.

② 위원회는 위원장 2명을 포함한 50명 이내의 위원으로 구성한다.

③ 위원회의 위원장은 국무총리와 제4항 제2호의 위원 중에서

대통령이 지명하는 사람이 된다.

④ 위원회의 위원은 다음 각 호의 사람이 된다.

1. 기획재정부장관, 교육과학기술부장관, 지식경제부장관, 환경부장관, 국토해양부장관 등 대통령령으로 정하는 공무원
2. 기후변화, 에너지·자원, 녹색기술·녹색산업, 지속가능발전 분야 등 저탄소 녹색성장에 관한 학식과 경험이 풍부한 사람 중에서 대통령이 위촉하는 사람

⑤ 위원회의 사무를 처리하게 하기 위하여 위원회에 간사위원 1명을 두며, 간사위원의 지명에 관한 사항은 대통령령으로 정한다.

⑥ 위원장은 각자 위원회를 대표하며, 위원회의 업무를 총괄한다.

⑦ 위원장이 부득이한 사유로 직무를 수행할 수 없는 때에는 국무총리인 위원장이 미리 정한 위원이 위원장의 직무를 대행한다.

⑧ 제4항 제2호의 위원의 임기는 1년으로 하되, 연임할 수 있다.

제15조【위원회의 기능】 위원회는 다음 각 호의 사항을 심의한다.

1. 저탄소 녹색성장 정책의 기본방향에 관한 사항
2. 녹색성장국가전략의 수립·변경·시행에 관한 사항
3. 기후변화대응 기본계획, 에너지기본계획 및 지속가능발전 기본계획에 관한 사항
4. 저탄소 녹색성장 추진의 목표 관리, 점검, 실태조사 및 평가에 관한 사항
5. 관계 중앙행정기관 및 지방자치단체의 저탄소 녹색성장과 관련된 정책 조정 및 지원에 관한 사항
6. 저탄소 녹색성장과 관련된 법제도에 관한 사항
7. 저탄소 녹색성장을 위한 재원의 배분방향 및 효율적 사용에 관한 사항
8. 저탄소 녹색성장과 관련된 국제협상·국제협력, 교육·홍

보, 인력양성 및 기반구축 등에 관한 사항

9. 저탄소 녹색성장과 관련된 기업 등의 고충조사, 처리, 시정권고 또는 의견표명

10. 다른 법률에서 위원회의 심의를 거치도록 한 사항

11. 그 밖에 저탄소 녹색성장과 관련하여 위원장이 필요하다고 인정하는 사항

제16조【회의】 ① 위원장은 위원회의 회의를 소집하고 그 의장이 된다.

② 위원회의 회의는 정기회의와 임시회의로 구분하며, 임시회의는 위원장이 필요하다고 인정하는 경우 또는 위원 5명 이상의 소집요구가 있을 경우에 위원장이 소집한다.

③ 위원회의 회의는 위원 과반수의 출석으로 개의하고, 출석위원 과반수의 찬성으로 의결한다. 다만, 대통령령으로 정하는 경우에는 서면으로 심의·의결할 수 있다.

④ 제1항부터 제3항까지에서 규정한 사항 외에 정기회의의 시기 등 위원회의 운영에 필요한 사항은 대통령령으로 정한다.

제17조【분과위원회】 ① 위원회의 업무를 효율적으로 수행·지원하고 위원회가 위임하는 업무를 검토·조정 또는 처리하기 위하여 대통령령으로 정하는 바에 따라 위원회에 분과위원회를 둘 수 있다.

② 분과위원회는 위촉위원으로 구성하며, 분과위원회의 위원장은 분과위원회의 위원 중에서 호선(互選)한다.

③ 중앙행정기관의 고위공무원단에 속하는 공무원은 관계 분야의 안건에 대하여 해당 분과위원회에 참석하여 의견을 제시할 수 있다.

④ 제1항부터 제3항까지에서 규정한 사항 외에 분과위원회의 운영에 필요한 사항은 위원회의 의결을 거쳐 위원회의 위원장이 정한다.

제18조【녹색성장기획단】 ① 위원회 및 분과위원회의 운영 및 업무를 효율적으로 지원하기 위하여 위원회에 녹색성장기획단(이하 "기획단"이라 한다)을 둔다.

② 기획단의 구성 및 운영 등에 필요한 사항은 대통령령으로 정한다.

제19조【공무원 등의 파견 요청】 위원회는 위원회의 운영 또는 기획단의 업무수행을 위하여 필요한 경우에는 중앙행정기관, 지방자치단체 소속의 공무원 및 관련 민간기관·단체 또는 연구소, 기업 임직원 등의 파견 또는 겸임을 요청할 수 있다.

제20조【지방녹색성장위원회의 구성 및 운영】 ① 지방자치단체의 저탄소 녹색성장과 관련된 주요 정책 및 계획과 그 이행에 관한 사항을 심의하기 위하여 시·도지사 소속으로 지방녹색성장위원회(이하 "지방녹색성장위원회"라 한다)를 둘 수 있다.

② 지방녹색성장위원회의 구성, 운영 및 기능 등에 필요한 사항은 대통령령으로 정한다.

제21조【녹색성장책임관의 지정】 저탄소 녹색성장의 원활한 추진을 위하여 중앙행정기관의 장 및 시·도지사는 소속 공무원 중에서 녹색성장책임관을 지정할 수 있다.

제 4 장 저탄소 녹색성장의 추진

제22조【녹색경제·녹색산업 구현을 위한 기본원칙】 ① 정부는 화석연료의 사용을 단계적으로 축소하고 녹색기술과 녹색산업을 육성함으로써 국가경쟁력을 강화하고 지속가능발전을 추구하는 경제(이하 "녹색경제"라 한다)를 구현하여야 한다.

② 정부는 녹색경제 정책을 수립·시행할 때 금융·산업·과학기술·환경·국토·문화 등 다양한 부문을 통합적 관점에서 균형 있게 고려하여야 한다.

③ 정부는 새로운 녹색산업의 창출, 기존 산업의 녹색산업으로의 전환 및 관련 산업과의 연계 등을 통하여 에너지·자원 다소비형 산업구조가 저탄소 녹색산업구조로 단계적으로 전환되도록 노력하여야 한다.

④ 정부는 저탄소 녹색성장을 추진할 때 지역 간 균형발전을 도모하며 저소득층이 소외되지 않도록 지원 및 배려하여야 한다.

제23조【녹색경제·녹색산업의 육성·지원】 ① 정부는 녹색경제를 구현함으로써 국가경제의 건전성과 경쟁력을 강화하고 성장잠재력이 큰 새로운 녹색산업을 발굴·육성하는 등 녹색경제·녹색산업의 육성·지원 시책을 마련하여야 한다.

② 제1항에 따른 녹색경제·녹색산업의 육성·지원 시책에는 다음 각 호의 사항이 포함되어야 한다.

1. 국내외 경제여건 및 전망에 관한 사항
2. 기존 산업의 녹색산업 구조로의 단계적 전환에 관한 사항
3. 녹색산업을 촉진하기 위한 중장기·단계별 목표, 추진전략에 관한 사항
4. 녹색산업의 신성장동력으로의 육성·지원에 관한 사항
5. 전기·정보통신·교통시설 등 기존 국가기반시설의 친환경 구조로의 전환에 관한 사항
6. 녹색경영을 위한 자문서비스 산업의 육성에 관한 사항
7. 녹색산업 인력 양성 및 일자리 창출에 관한 사항
8. 그 밖에 녹색경제·녹색산업의 촉진에 관한 사항

제24조【자원순환의 촉진】 ① 정부는 자원을 절약하고 효율적으로 이용하며 폐기물의 발생을 줄이는 등 자원순환의 촉진과 자원생산성 제고를 위하여 자원순환 산업을 육성·지원하기 위한

다양한 시책을 마련하여야 한다.

② 제1항에 따른 자원순환 산업의 육성·지원 시책에는 다음 각 호의 사항이 포함되어야 한다.

1. 자원순환 촉진 및 자원생산성 제고 목표설정
2. 자원의 수급 및 관리
3. 유해하거나 재제조·재활용이 어려운 물질의 사용억제
4. 폐기물 발생의 억제 및 재제조·재활용 등 재자원화
5. 에너지자원으로 이용되는 목재, 식물, 농산물 등 바이오매스의 수집·활용
6. 자원순환 관련 기술개발 및 산업의 육성
7. 자원생산성 향상을 위한 교육훈련·인력양성 등에 관한 사항

제25조【기업의 녹색경영 촉진】 ① 정부는 기업의 녹색경영을 지원·촉진하여야 한다.

② 정부는 기업의 녹색경영을 지원·촉진하기 위하여 다음 각 호의 사항을 포함하는 시책을 수립·시행하여야 한다.

1. 친환경 생산체제로의 전환을 위한 기술지원
2. 기업의 에너지·자원 이용 효율화, 온실가스 배출량 감축, 산림조성 및 자연환경 보전, 지속가능발전 정보 등 녹색경영 성과의 공개
3. 중소기업의 녹색경영에 대한 지원
4. 그 밖에 저탄소 녹색성장을 위한 기업활동 지원에 관한 사항

제26조【녹색기술의 연구개발 및 사업화 등의 촉진】 ① 정부는 녹색기술의 연구개발 및 사업화 등을 촉진하기 위하여 다음 각 호의 사항을 포함하는 시책을 수립·시행할 수 있다.

1. 녹색기술과 관련된 정보의 수집·분석 및 제공
2. 녹색기술 평가기법의 개발 및 보급

3. 녹색기술 연구개발 및 사업화 등의 촉진을 위한 금융지원
4. 녹색기술 전문인력의 양성 및 국제협력 등

② 정부는 정보통신・나노・생명공학 기술 등의 융합을 촉진하고 녹색기술의 지식재산권화를 통하여 저탄소 지식기반경제로의 이행을 신속하게 추진하여야 한다.

③ 「과학기술기본법」에 따른 과학기술기본계획에 제1항의 시책이 포함되는 경우에는 미리 위원회의 의견을 들어야 한다.

제27조【정보통신기술의 보급・활용】 ① 정부는 에너지 절약, 에너지 이용효율 향상 및 온실가스 감축을 위하여 정보통신기술 및 서비스를 적극 활용하는 다음 각 호에 대한 시책을 수립・시행하여야 한다.

1. 방송통신 네트워크 등 정보통신 기반 확대
2. 새로운 정보통신 서비스의 개발・보급
3. 정보통신 산업 및 기기 등에 대한 녹색기술 개발 촉진

② 정부는 저탄소 녹색성장을 위한 생활문화를 조속히 확산시키기 위하여 재택근무・영상회의・원격교육・원격진료 등을 활성화하는 등의 방송통신 시책을 수립・시행하여야 한다.

③ 정부는 정보통신기술을 활용하여 전력 네트워크를 지능화・고도화함으로써 고품질의 전력서비스를 제공하고 에너지 이용효율을 극대화하며 온실가스를 획기적으로 감축할 수 있도록 하여야 한다.

제28조【금융의 지원 및 활성화】 정부는 저탄소 녹색성장을 촉진하기 위하여 다음 각 호의 사항을 포함하는 금융 시책을 수립・시행하여야 한다.

1. 녹색경제 및 녹색산업의 지원 등을 위한 재원의 조성 및 자금 지원
2. 저탄소 녹색성장을 지원하는 새로운 금융상품의 개발
3. 저탄소 녹색성장을 위한 기반시설 구축사업에 대한 민간

투자 활성화

4. 기업의 녹색경영 정보에 대한 공시제도 등의 강화 및 녹색경영 기업에 대한 금융지원 확대

5. 탄소시장(온실가스를 배출할 수 있는 권리 또는 온실가스의 감축·흡수 실적 등을 거래하는 시장을 말한다. 이하 같다)의 개설 및 거래 활성화 등

제29조【녹색산업투자회사의 설립과 지원】 ① 녹색기술 및 녹색산업에 자산을 투자하여 그 수익을 투자자에게 배분하는 것을 목적으로 하는 녹색산업투자회사(「자본시장과 금융투자업에 관한 법률」 제9조 제18항의 집합투자기구를 말한다. 이하 같다)를 설립할 수 있다.

② 녹색산업투자회사가 투자하는 녹색기술 및 녹색산업은 다음 각호에서 정하는 사업 또는 기업으로 한다.

1. 제2조 제3호에 따른 녹색기술에 대한 연구와 시제품의 제작 및 상용화를 위한 연구개발 또는 기술지원 사업
2. 제2조 제4호에 따른 녹색산업에 해당하는 사업
3. 녹색기술 또는 녹색산업에 대한 투자 또는 영업을 영위하는 기업

③ 정부는 「공공기관의 운영에 관한 법률」 제4조에 따른 공공기관이 녹색산업투자회사에 출자하려는 경우 이를 위한 자금의 전부 또는 일부를 예산의 범위에서 지원할 수 있다.

④ 금융위원회는 제3항의 규정에 따라 공공기관이 출자한 녹색산업투자회사(해당 회사의 자산운용회사·자산보관회사 및 일반사무관리회사를 포함한다. 이하 이 조에서 같다)에게 해당 회사의 업무 및 재산 등에 관한 자료의 제출이나 보고를 요구할 수 있으며, 관계 중앙행정기관은 금융위원회에 해당 자료의 제출을 요구할 수 있다.

⑤ 관계 중앙행정기관은 제4항에 의하여 제출된 자료나 보고

내용에 대하여 검사가 필요하다고 인정하는 경우 금융위원회에게 해당 녹색산업투자회사에 대한 업무 및 재산 등에 관한 검사를 요청할 수 있으며, 해당 검사 결과 중대한 문제가 있다고 여겨지는 경우에는 금융위원회는 관계 중앙행정기관과 협의하여 해당 녹색산업투자회사의 등록을 취소할 수 있다.

⑥ 제1항 내지 제5항에 따른 녹색산업투자회사의 설립·운영 및 재정지원과 그 밖에 필요한 세부사항은 대통령령으로 정한다.

제30조【조세 제도 운영】 정부는 에너지·자원의 위기 및 기후변화 문제에 효과적으로 대응하고 저탄소 녹색성장을 촉진하기 위하여 온실가스와 오염물질을 발생시키거나 에너지·자원 이용효율이 낮은 재화와 서비스를 줄이고 환경친화적인 재화와 서비스를 촉진하는 방향으로 국가의 조세 제도를 운영하여야 한다.

제31조【녹색기술·녹색산업에 대한 지원·특례 등】 ① 국가 또는 지방자치단체는 녹색기술·녹색산업에 대하여 보조금의 지급 등 필요한 지원을 할 수 있다.

② 「신용보증기금법」에 따라 설립된 신용보증기금 및 「기술신용보증기금법」에 따라 설립된 기술신용보증기금은 녹색기술·녹색산업에 우선적으로 신용보증을 하거나 보증조건 등을 우대할 수 있다.

③ 국가나 지방자치단체는 녹색기술·녹색산업과 관련된 기업을 지원하기 위하여 「조세특례제한법」과 「지방세법」에서 정하는 바에 따라 소득세·법인세·취득세·재산세·등록세 등을 감면할 수 있다.

④ 국가나 지방자치단체는 녹색기술·녹색산업과 관련된 기업이 「외국인투자 촉진법」 제2조 제1항 제4호에 따른 외국인투자를 유치하는 경우에 이를 최대한 지원하기 위하여 노력하여야 한다.

제32조【녹색기술 · 녹색산업의 표준화 및 인증 등】 ① 정부는 국내에서 개발되었거나 개발 중인 녹색기술 · 녹색산업이 「국가표준기본법」 제3조 제2호에 따른 국제표준에 부합되도록 표준화 기반을 구축하고 녹색기술 · 녹색산업의 국제표준화 활동 등에 필요한 지원을 할 수 있다.

② 정부는 녹색기술 · 녹색산업의 발전을 촉진하기 위하여 녹색기술, 녹색사업, 녹색제품 등에 대한 적합성 인증을 하거나 녹색전문기업 확인, 공공기관의 구매의무화 또는 기술지도 등을 할 수 있다.

③ 정부는 다음 각 호의 어느 하나에 해당하는 경우에는 제2항에 따른 적합성 인증 및 녹색전문기업 확인을 취소하여야 한다.

1. 거짓이나 그 밖의 부정한 방법으로 인증이나 확인을 받은 경우
2. 중대한 결함이 있어 인증이나 확인이 적당하지 아니하다고 인정되는 경우

④ 제1항 내지 제3항에 따른 표준화, 인증 및 취소 등에 관하여 그 밖에 필요한 사항은 대통령령으로 정한다.

제33조【중소기업의 지원 등】 정부는 중소기업의 녹색기술 및 녹색경영을 촉진하기 위하여 다음 각 호의 시책을 수립 · 시행할 수 있다.

1. 대기업과 중소기업의 공동사업에 대한 우선 지원
2. 대기업의 중소기업에 대한 기술지도 · 기술이전 및 기술인력 파견에 대한 지원
3. 중소기업의 녹색기술 사업화의 촉진
4. 녹색기술 개발 촉진을 위한 공공시설의 이용
5. 녹색기술 · 녹색산업에 관한 전문인력 양성 · 공급 및 국외진출
6. 그 밖에 중소기업의 녹색기술 및 녹색경영을 촉진하기

위한 사항

제34조【녹색기술 · 녹색산업 집적지 및 단지 조성 등】 ① 정부는 녹색기술의 공동연구개발, 시설장비의 공동활용 및 산 · 학 · 연 네트워크 구축 등의 사업을 위한 집적지와 단지를 조성하거나 이를 지원할 수 있다.

② 제1항에 따른 사업을 추진하는 경우에는 다음 각 호의 사항을 고려하여야 한다.

1. 산업단지별 산업집적 현황에 관한 사항
2. 기업 · 대학 · 연구소 등의 연구개발 역량강화 및 상호연계에 관한 사항
3. 산업집적기반시설의 확충 및 우수한 녹색기술 · 녹색산업 인력의 유치에 관한 사항
4. 녹색기술 · 녹색산업의 사업추진체계 및 재원조달방안

③ 정부는 대통령령으로 정하는 기관 또는 단체로 하여금 녹색기술 · 녹색산업 집적지 및 단지를 조성하게 할 수 있다.

④ 정부는 제3항에 따른 기관 또는 단체가 같은 항에 따른 녹색기술 · 녹색산업 집적지 및 단지를 조성하는 사업을 수행하는 데에 소요되는 비용의 전부 또는 일부를 출연할 수 있다.

제35조【녹색기술 · 녹색산업에 대한 일자리 창출 등】 ① 정부는 녹색기술 · 녹색산업에 대한 일자리를 창출 · 확대하여 모든 국민이 녹색성장의 혜택을 누릴 수 있도록 하여야 한다.

② 정부는 녹색기술 · 녹색산업에 대한 일자리를 창출하는 과정에서 산업분야별 노동력의 원활한 이동 · 전환을 촉진하고 국민이 새로운 기술을 습득할 수 있는 기회를 확대하며, 녹색기술 · 녹색산업에 대한 일자리 창출을 위한 재정적 · 기술적 지원을 할 수 있다.

제36조【규제의 선진화】 ① 정부는 자원을 효율적으로 이용하고

온실가스와 오염물질의 발생을 줄이기 위한 규제를 도입하려는 경우에는 온실가스 또는 오염물질의 발생 원인자가 스스로 온실가스와 오염물질의 발생을 줄이도록 유도함으로써 사회·경제적 비용을 줄이도록 노력하여야 한다.

② 정부는 온실가스와 오염물질의 발생을 줄이기 위한 규제를 도입하려는 경우에는 민간의 자율과 창의를 저해하지 않도록 하고, 기업의 규제에 대한 국내외 실태조사 등을 하여 산업경쟁력을 높일 수 있도록 규제의 중복을 피하는 등 규제 체계를 선진화하여야 한다.

제37조【국제규범 대응】 ① 정부는 외국 정부 또는 국제기구에서 제정하거나 도입하려는 저탄소 녹색성장과 관련된 제도·정책에 관한 동향과 정보를 수집·조사·분석하여 관련 제도·정책을 합리적으로 정비하고 지원체제를 구축하는 등 적절한 대책을 마련하여야 한다.

② 정부는 제1항의 동향·정보 및 대책에 관한 사항을 기업·국민들에게 충분히 제공함으로써 국내 기업과 국민들이 대응역량을 높일 수 있도록 하여야 한다.

제 5 장 저탄소 사회의 구현

제38조【기후변화대응의 기본원칙】 정부는 저탄소 사회를 구현하기 위하여 기후변화대응 정책 및 관련 계획을 다음 각 호의 원칙에 따라 수립·시행하여야 한다.

1. 지구온난화에 따른 기후변화 문제의 심각성을 인식하고 국가적·국민적 역량을 모아 총체적으로 대응하고 범지구적 노력에 적극 참여한다.

2. 온실가스 감축의 비용과 편익을 경제적으로 분석하고 국내 여건 등을 감안하여 국가온실가스 중장기 감축 목표를 설정하고, 가격기능과 시장원리에 기반을 둔 비용효과적 방식의 합리적 규제체제를 도입함으로써 온실가스 감축을 효율적·체계적으로 추진한다.
3. 온실가스를 획기적으로 감축하기 위하여 정보통신·나노·생명 공학 등 첨단기술 및 융합기술을 적극 개발하고 활용한다.
4. 온실가스 배출에 따른 권리·의무를 명확히 하고 이에 대한 시장거래를 허용함으로써 다양한 감축수단을 자율적으로 선택할 수 있도록 하고, 국내 탄소시장을 활성화하여 국제 탄소시장에 적극 대비한다.
5. 대규모 자연재해, 환경생태와 작물상황의 변화에 대비하는 등 기후변화로 인한 영향을 최소화하고 그 위험 및 재난으로부터 국민의 안전과 재산을 보호한다.

제39조【에너지정책 등의 기본원칙】 정부는 저탄소 녹색성장을 추진하기 위하여 에너지정책 및 에너지와 관련된 계획을 다음 각 호의 원칙에 따라 수립·시행하여야 한다.

1. 석유·석탄 등 화석연료의 사용을 단계적으로 축소하고 에너지 자립도를 획기적으로 향상시킨다.
2. 에너지 가격의 합리화, 에너지의 절약, 에너지 이용효율 제고 등 에너지 수요관리를 강화하여 지구온난화를 예방하고 환경을 보전하며, 에너지 저소비·자원순환형 경제·사회구조로 전환한다.
3. 친환경에너지인 태양에너지, 폐기물·바이오에너지, 풍력, 지열, 조력, 연료전지, 수소에너지 등 신·재생에너지의 개발·생산·이용 및 보급을 확대하고 에너지 공급원을 다변화한다.

4. 에너지가격 및 에너지산업에 대한 시장경쟁 요소의 도입을 확대하고 공정거래 질서를 확립하며, 국제규범 및 외국의 법제도 등을 고려하여 에너지산업에 대한 규제를 합리적으로 도입·개선하여 새로운 시장을 창출한다.
5. 국민이 저탄소 녹색성장의 혜택을 고루 누릴 수 있도록 저소득층에 대한 에너지 이용 혜택을 확대하고 형평성을 제고하는 등 에너지와 관련한 복지를 확대한다.
6. 국외 에너지자원 확보, 에너지의 수입 다변화, 에너지 비축 등을 통하여 에너지를 안정적으로 공급함으로써 에너지에 관한 국가안보를 강화한다.

제40조【기후변화대응 기본계획】 ① 정부는 기후변화대응의 기본원칙에 따라 20년을 계획기간으로 하는 기후변화대응 기본계획을 5년마다 수립·시행하여야 한다.

② 기후변화대응 기본계획을 수립하거나 변경하는 경우에는 위원회의 심의 및 국무회의 심의를 거쳐야 한다. 다만, 대통령령으로 정하는 경미한 사항을 변경하는 경우에는 그러하지 아니하다.

③ 기후변화대응 기본계획에는 다음 각 호의 사항이 포함되어야 한다.

1. 국내외 기후변화 경향 및 미래 전망과 대기 중의 온실가스 농도변화
2. 온실가스 배출·흡수 현황 및 전망
3. 온실가스 배출 중장기 감축목표 설정 및 부문별·단계별 대책
4. 기후변화대응을 위한 국제협력에 관한 사항
5. 기후변화대응을 위한 국가와 지방자치단체의 협력에 관한 사항
6. 기후변화대응 연구개발에 관한 사항

7. 기후변화대응 인력양성에 관한 사항
8. 기후변화의 감시 · 예측 · 영향 · 취약성평가 및 재난방지 등 적응대책에 관한 사항
9. 기후변화대응을 위한 교육 · 홍보에 관한 사항
10. 그 밖에 기후변화대응 추진을 위하여 필요한 사항

제41조【에너지기본계획의 수립】 ① 정부는 에너지정책의 기본원칙에 따라 20년을 계획기간으로 하는 에너지기본계획(이하 이 조에서 “에너지기본계획”이라 한다)을 5년마다 수립 · 시행하여야 한다.

② 에너지기본계획을 수립하거나 변경하는 경우에는 「에너지법」 제9조에 따른 에너지위원회의 심의를 거친 다음 위원회와 국무회의의 심의를 거쳐야 한다. 다만, 대통령령으로 정하는 경미한 사항을 변경하는 경우에는 그러하지 아니하다.

③ 에너지기본계획에는 다음 각 호의 사항이 포함되어야 한다.

1. 국내외 에너지 수요와 공급의 추이 및 전망에 관한 사항
2. 에너지의 안정적 확보, 도입 · 공급 및 관리를 위한 대책에 관한 사항
3. 에너지 수요 목표, 에너지원 구성, 에너지 절약 및 에너지 이용효율 향상에 관한 사항
4. 신 · 재생에너지 등 환경친화적 에너지의 공급 및 사용을 위한 대책에 관한 사항
5. 에너지 안전관리를 위한 대책에 관한 사항
6. 에너지 관련 기술개발 및 보급, 전문인력 양성, 국제협력, 부존 에너지자원 개발 및 이용, 에너지 복지 등에 관한 사항

제42조【기후변화대응 및 에너지의 목표관리】 ① 정부는 범지구적인 온실가스 감축에 적극 대응하고 저탄소 녹색성장을 효율적 · 체계적으로 추진하기 위하여 다음 각 호의 사항에 대한 중

장기 및 단계별 목표를 설정하고 그 달성을 위하여 필요한 조치를 강구하여야 한다.

1. 온실가스 감축 목표
2. 에너지 절약 목표 및 에너지 이용효율 목표
3. 에너지 자립 목표
4. 신·재생에너지 보급 목표

② 정부는 제1항에 따른 목표를 설정할 때 국내 여건 및 각국의 동향 등을 고려하여야 한다.

③ 정부는 제1항에 따른 목표를 달성하기 위하여 관계 중앙행정기관, 지방자치단체 및 대통령령으로 정하는 공공기관 등에 대하여 대통령령으로 정하는 바에 따라 해당 기관별로 에너지 절약 및 온실가스 감축목표를 설정하도록 하고 그 이행사항을 지도·감독할 수 있다.

④ 정부는 제1항 제1호 및 제2호에 따른 목표를 달성할 수 있도록 산업, 교통·수송, 가정·상업 등 부문별 목표를 설정하고 그 달성을 위하여 필요한 조치를 적극 마련하여야 한다.

⑤ 정부는 제1항 제1호 및 제2호에 따른 목표를 달성하기 위하여 대통령령으로 정하는 기준량 이상의 온실가스 배출업체 및 에너지 소비업체(이하 "관리업체"라 한다)별로 측정·보고·검증이 가능한 방식으로 목표를 설정·관리하여야 한다. 이 경우 정부는 관리업체와 미리 협의하여야 하며, 온실가스 배출 및 에너지 사용 등의 이력, 기술 수준, 국제경쟁력, 국가목표 등을 고려하여야 한다.

⑥ 관리업체는 제5항에 따른 목표를 준수하여야 하며, 그 실적을 대통령령으로 정하는 바에 따라 정부에 보고하여야 한다.

⑦ 정부는 제6항에 따라 보고받은 실적에 대하여 등록부를 작성하고 체계적으로 관리하여야 한다.

⑧ 정부는 관리업체의 준수실적이 제5항에 따른 목표에 미달하는 경우 목표달성을 위하여 필요한 개선을 명할 수 있다. 이

경우 관리업체는 개선명령에 따른 이행계획을 작성하여 이를 성실히 이행하여야 한다.

⑨ 관리업체는 제8항에 따른 이행결과를 측정·보고·검증이 가능한 방식으로 작성하여 대통령령으로 정하는 공신력 있는 외부 전문기관의 검증을 받아 정부에 보고하고 공개하여야 한다.

⑩ 정부는 관리업체가 제5항에 따른 목표를 달성하고 제8항에 따른 이행계획을 차질 없이 이행할 수 있도록 하기 위하여 필요한 경우 재정·세제·경영·기술지원, 실태조사 및 진단, 자료 및 정보의 제공 등을 할 수 있다.

⑪ 제5항부터 제9항까지에서 규정한 사항 외에 등록부의 관리, 관리업체의 지원 등에 필요한 사항은 대통령령으로 정한다.

제43조【온실가스 감축의 조기행동 촉진】 ① 정부는 관리업체가 제42조 제5항에 따른 목표관리를 받기 전에 자발적으로 행한 실적에 대해서는 이를 목표관리 실적으로 인정하거나 그 실적을 거래할 수 있도록 하는 등 자발적으로 온실가스를 미리 감축하는 행동을 하도록 촉진하여야 한다.

② 제1항에 따른 실적을 거래할 수 있는 방법 및 절차 등에 필요한 사항은 대통령령으로 정한다.

제44조【온실가스 배출량 및 에너지 사용량 등의 보고】 ① 관리업체는 사업장별로 매년 온실가스 배출량 및 에너지 소비량에 대하여 측정·보고·검증 가능한 방식으로 명세서를 작성하여 정부에 보고하여야 한다.

② 관리업체는 제1항에 따른 보고를 할 때 명세서의 신뢰성 여부에 대하여 대통령령으로 정하는 공신력 있는 외부 전문기관의 검증을 받아야 한다. 이 경우 정부는 명세서에 흠이 있거나 빠진 부분에 대하여 시정 또는 보완을 명할 수 있다.

③ 정부는 명세서를 체계적으로 관리하고 명세서에 포함된 주요 정보를 관리업체별로 공개할 수 있다. 다만, 관리업체는 정

보공개로 인하여 그 관리업체의 권리나 영업상의 비밀이 현저히 침해되는 특별한 사유가 있는 경우에는 비공개를 요청할 수 있다.

④ 정부는 관리업체로부터 제3항 단서에 따른 정보의 비공개 요청을 받았을 때에는 심사위원회를 구성하여 30일 이내에 그 결과를 통지하여야 한다.

⑤ 명세서의 내용, 보고·관리, 공개방법 및 심사위원회의 구성·운영 등에 필요한 사항은 대통령령으로 정한다.

제45조【온실가스 종합정보관리체계의 구축】 ① 정부는 국가 온실가스 배출량·흡수량, 배출·흡수 계수(係數), 온실가스 관련 각종 정보 및 통계를 개발·검증·관리하는 온실가스 종합정보관리체계를 구축하여야 한다.

② 관계 중앙행정기관의 장은 제1항에 따른 종합정보관리체계가 원활히 운영될 수 있도록 에너지·산업공정·농업·폐기물·산림 등 부문별 소관 분야의 정보 및 통계를 작성하여 제공하는 등 적극 협력하여야 한다.

③ 정부는 제1항에 따른 각종 정보 및 통계를 작성·관리하거나 종합정보관리체계를 구축함에 있어 국제기준을 최대한 반영하여 전문성·투명성 및 신뢰성을 제고하여야 한다.

④ 정부는 제1항에 따른 각종 정보 및 통계를 분석·검증하여 그 결과를 매년 공표하여야 한다.

⑤ 제1항부터 제4항까지에서 규정한 사항 외에 세부적인 정보 및 통계 관리방법, 관리기관 및 방법 등은 대통령령으로 정한다.

제46조【총량제한 배출권 거래제 등의 도입】 ① 정부는 시장기능을 활용하여 효율적으로 국가의 온실가스 감축목표를 달성하기 위하여 온실가스 배출권을 거래하는 제도를 운영할 수 있다.

② 제1항의 제도에는 온실가스 배출허용총량을 설정하고 배출권을 거래하는 제도 및 기타 국제적으로 인정되는 거래 제도를

포함한다.

③ 정부는 제2항에 따른 제도를 실시할 경우 기후변화 관련 국제협상을 고려하여야 하고, 국제경쟁력이 현저하게 약화될 우려가 있는 제42조 제5항의 관리업체에 대하여는 필요한 조치를 강구할 수 있다.

④ 제2항에 따른 제도의 실시를 위한 배출허용량의 할당방법, 등록·관리방법 및 거래소 설치·운영 등은 따로 법률로 정한다.

제47조【교통부문의 온실가스 관리】 ① 자동차 등 교통수단을 제작하려는 자는 그 교통수단에서 배출되는 온실가스를 감축하기 위한 방안을 마련하여야 하며, 온실가스 감축을 위한 국제경쟁체제에 부응할 수 있도록 적극 노력하여야 한다.

② 정부는 자동차의 평균에너지소비효율을 개선함으로써 에너지 절약을 도모하고, 자동차 배기가스 중 온실가스를 줄임으로써 쾌적하고 적정한 대기환경을 유지할 수 있도록 자동차 평균에너지소비효율기준 및 자동차 온실가스 배출허용기준을 각각 정하되, 이중규제가 되지 않도록 자동차 제작업체(수입업체를 포함한다)로 하여금 어느 한 기준을 택하여 준수토록 하고 측정방법 등이 중복되지 않도록 하여야 한다.

③ 정부는 온실가스 배출량이 적은 자동차 등을 구매하는 자에 대하여 재정적 지원을 강화하고 온실가스 배출량이 많은 자동차 등을 구매하는 자에 대해서는 부담금을 부과하는 등의 방안을 강구할 수 있다.

④ 정부는 하이브리드 자동차, 수소연료전지 자동차 등 저탄소·고효율 교통수단의 제작·보급을 촉진하기 위하여 재정·세제 지원, 연구개발 및 관련 제도 개선 등의 방안을 강구할 수 있다.

제48조【기후변화 영향평가 및 적응대책의 추진】 ① 정부는 기상현상에 대한 관측·예측·제공·활용 능력을 높이고, 지역별·

권역별로 태양력·풍력·조력 등 신·재생에너지원을 확보할 수 있는 잠재력을 지속적으로 분석·평가하여 이에 관한 기상정보관리체계를 구축·운영하여야 한다.
② 정부는 기후변화에 대한 감시·예측의 정확도를 향상시키고 생물자원 및 수자원 등의 변화 상황과 국민건강에 미치는 영향 등 기후변화로 인한 영향을 조사·분석하기 위한 조사·연구, 기술개발, 관련 전문기관의 지원 및 국내외 협조체계 구축 등의 시책을 추진하여야 한다.
③ 정부는 관계 중앙행정기관의 장과 협의하여 기후변화로 인한 생태계, 생물다양성, 대기, 수자원·수질, 보건, 농·수산식품, 산림, 해양, 산업, 방재 등에 미치는 영향 및 취약성을 조사·평가하고 그 결과를 공표하여야 한다.
④ 정부는 기후변화로 인한 피해를 줄이기 위하여 사전 예방적 관리에 우선적인 노력을 기울여야 하며 대통령령으로 정하는 바에 따라 기후변화의 영향을 완화시키거나 건강·자연재해 등에 대응하는 적응대책을 수립·시행하여야 한다.
⑤ 정부는 국민·사업자 등이 기후변화 적응대책에 따라 활동할 경우 이에 필요한 기술적 및 재정적 지원을 할 수 있다.

제 6 장 녹색생활 및 지속가능발전의 실현

제49조【녹색생활 및 지속가능발전의 기본원칙】 녹색생활 및 지속가능발전의 실현을 위한 국가의 시책은 다음 각 호의 기본원칙에 따라 추진되어야 한다.

1. 국토는 녹색성장의 터전이며 그 결과의 전시장이라는 점을 인식하고 현세대 및 미래세대가 쾌적한 삶을 영위할 수 있도록 국토의 개발 및 보전·관리가 조화될 수 있도

록 한다.

2. 국토·도시공간구조와 건축·교통체제를 저탄소 녹색성장 구조로 개편하고 생산자와 소비자가 녹색제품을 자발적·적극적으로 생산하고 구매할 수 있는 여건을 조성한다.
3. 국가·지방자치단체·기업 및 국민은 지속가능발전과 관련된 국제적 합의를 성실히 이행하고, 국민의 일상생활 속에 녹색생활이 내재화되고 녹색문화가 사회전반에 정착될 수 있도록 한다.
4. 국가·지방자치단체 및 기업은 경제발전의 기초가 되는 생태학적 기반을 보호할 수 있도록 토지이용과 생산시스템을 개발·정비함으로써 환경보전을 촉진한다.

제50조【지속가능발전 기본계획의 수립·시행】 ① 정부는 1992년 브라질에서 개최된 유엔환경개발회의에서 채택한 의제21, 2002년 남아프리카공화국에서 개최된 세계지속가능발전정상회의에서 채택한 이행계획 등 지속가능발전과 관련된 국제적 합의를 성실히 이행하고, 국가의 지속가능발전을 촉진하기 위하여 20년을 계획기간으로 하는 지속가능발전 기본계획을 5년마다 수립·시행하여야 한다.

② 지속가능발전 기본계획을 수립하거나 변경하는 경우에는 「지속가능발전법」 제15조에 따른 지속가능발전위원회의 심의를 거친 다음 위원회와 국무회의의 심의를 거쳐야 한다. 다만, 대통령령으로 정하는 경미한 사항을 변경하는 경우에는 그러하지 아니하다.

③ 지속가능발전 기본계획에는 다음 각 호의 사항이 포함되어야 한다.

1. 지속가능발전의 현황 및 여건변화와 전망에 관한 사항
2. 지속가능발전을 위한 비전, 목표, 추진전략과 원칙, 기본정책 방향, 주요지표에 관한 사항

3. 지속가능발전에 관련된 국제적 합의이행에 관한 사항
4. 그 밖에 지속가능발전을 위하여 필요한 사항

④ 중앙행정기관의 장은 제1항에 따른 지속가능발전 기본계획과 조화를 이루는 소관 분야의 중앙 지속가능발전 기본계획을 중앙추진계획에 포함하여 수립·시행하여야 한다.

⑤ 시·도지사는 제1항에 따른 지속가능발전 기본계획과 조화를 이루며 해당 지방자치단체의 지역적 특성과 여건을 고려한 지방 지속가능발전 기본계획을 지방추진계획에 포함하여 수립·시행하여야 한다.

제51조【녹색국토의 관리】 ① 정부는 건강하고 쾌적한 환경과 아름다운 경관이 경제발전 및 사회개발과 조화를 이루는 국토(이하 "녹색국토"라 한다)를 조성하기 위하여 국토종합계획·도시기본계획 등 대통령령으로 정하는 계획을 제49조에 따른 녹색생활 및 지속가능발전의 기본원칙에 따라 수립·시행하여야 한다.

② 정부는 녹색국토를 조성하기 위하여 다음 각 호의 사항을 포함하는 시책을 마련하여야 한다.

1. 에너지·자원 자립형 탄소중립도시 조성
2. 산림·녹지의 확충 및 광역생태축 보전
3. 해양의 친환경적 개발·이용·보존
4. 저탄소 항만의 건설 및 기존 항만의 저탄소 항만으로의 전환
5. 친환경 교통체계의 확충
6. 자연재해로 인한 국토 피해의 완화
7. 그 밖에 녹색국토 조성에 관한 사항

③ 정부는 「국토기본법」에 따른 국토종합계획, 「국가균형발전 특별법」에 따른 지역발전계획 등 대통령령으로 정하는 계획을 수립할 때에는 미리 위원회의 의견을 들어야 된다.

제52조【기후변화대응을 위한 물 관리】 정부는 기후변화로 인한

가뭄 등 자연재해와 물 부족 및 수질악화와 수생태계 변화에 효과적으로 대응하고 모든 국민이 물의 혜택을 고루 누릴 수 있도록 하기 위하여 다음 각 호의 사항을 포함하는 시책을 수립·시행하여야 한다.

1. 깨끗하고 안전한 먹는 물 공급과 가뭄 등에 대비한 안정적인 수자원의 확보
2. 수생태계의 보전·관리와 수질개선
3. 물 절약 등 수요관리, 빗물 이용·하수 재이용 등 순환체계의 정비 및 수해의 예방
4. 자연친화적인 하천의 보전·복원
5. 수질오염 예방·처리를 위한 기술 개발 및 관련 서비스 제공 등

제53조【저탄소 교통체계의 구축】 ① 정부는 교통부문의 온실가스 감축을 위한 환경을 조성하고 온실가스 배출 및 에너지의 효율적인 관리를 위하여 대통령령으로 정하는 바에 따라 온실가스 감축목표 등을 설정·관리하여야 한다.

② 정부는 에너지소비량과 온실가스 배출량을 최소화하는 저탄소 교통체계를 구축하기 위하여 대중교통분담률, 철도수송분담률 등에 대한 중장기 및 단계별 목표를 설정·관리하여야 한다.

③ 정부는 철도가 국가기간교통망의 근간이 되도록 철도에 대한 투자를 지속적으로 확대하고 버스·지하철·경전철 등 대중교통수단을 확대하며, 자전거 등의 이용 및 연안해운을 활성화하여야 한다.

④ 정부는 온실가스와 대기오염을 최소화하고 교통체증으로 인한 사회적 비용을 획기적으로 줄이며 대도시·수도권 등에서의 교통체증을 근본적으로 해결하기 위하여 다음 각 호의 사항을 포함하는 교통수요관리대책을 마련하여야 한다.

1. 혼잡통행료 및 교통유발부담금 제도 개선

2. 버스 · 저공해차량 전용차로 및 승용차진입제한 지역 확대
3. 통행량을 효율적으로 분산시킬 수 있는 지능형교통정보 시스템 확대 · 구축

제54조【녹색건축물의 확대】 ① 정부는 에너지이용 효율 및 신 · 재생에너지의 사용비율이 높고 온실가스 배출을 최소화하는 건축물(이하 "녹색건축물"이라 한다)을 확대하기 위하여 녹색건축물 등급제 등의 정책을 수립 · 시행하여야 한다.
② 정부는 건축물에 사용되는 에너지소비량과 온실가스 배출량을 줄이기 위하여 대통령령으로 정하는 기준 이상의 건물에 대한 중장기 및 기간별 목표를 설정 · 관리하여야 한다.
③ 정부는 건축물의 설계 · 건설 · 유지관리 · 해체 등의 전 과정에서 에너지 · 자원 소비를 최소화하고 온실가스 배출을 줄이기 위하여 설계기준 및 허가 · 심의를 강화하는 등 설계 · 건설 · 유지관리 · 해체 등의 단계별 대책 및 기준을 마련하여 시행하여야 한다.
④ 정부는 기존 건축물이 녹색건축물로 전환되도록 에너지 진단 및 「에너지이용 합리화법」 제25조에 따른 에너지절약사업과 이를 통한 온실가스 배출을 줄이는 사업을 지속적으로 추진하여야 한다.
⑤ 정부는 신축되거나 개축되는 건축물에 대해서는 전력소비량 등 에너지의 소비량을 조절 · 절약할 수 있는 지능형 계량기를 부착 · 관리하도록 할 수 있다.
⑥ 정부는 중앙행정기관, 지방자치단체, 대통령령으로 정하는 공공기관 및 교육기관 등의 건축물이 녹색건축물의 선도적 역할을 수행하도록 제1항부터 제5항까지의 규정에 따른 시책을 적용하고 그 이행사항을 점검 · 관리하여야 한다.
⑦ 정부는 대통령령으로 정하는 일정 규모 이상의 신도시의 개발 또는 도시 재개발을 하는 경우에는 녹색건축물을 확대 · 보

급하도록 노력하여야 한다.
⑧ 정부는 녹색건축물의 확대를 위하여 필요한 경우 대통령령으로 정하는 바에 따라 자금의 지원, 조세의 감면 등의 지원을 할 수 있다.

第55条【**친환경 농림수산의 촉진 및 탄소흡수원 확충**】 ① 정부는 에너지 절감 및 바이오에너지 생산을 위한 농업기술을 개발하고, 기후변화에 대응하는 친환경 농산물 생산기술을 개발하여 화학비료·자재와 농약사용을 최대한 억제하고 친환경·유기농 농수산물 및 나무제품의 생산·유통 및 소비를 확산하여야 한다.
② 정부는 농지의 보전·조성 및 바다숲(대기의 온실가스를 흡수하기 위하여 바다 속에 조성하는 우뭇가사리 등의 해조류군을 말한다)의 조성 등을 통하여 탄소흡수원을 확충하여야 한다.
③ 정부는 산림의 보전 및 조성을 통하여 탄소흡수원을 대폭 확충하고, 산림바이오매스 활용을 촉진하여야 한다.
④ 정부는 기후변화에 적극 대응할 수 있는 신품종 개량 등을 통하여 식량자립도를 높일 수 있는 시책을 수립·시행하여야 한다.

第56条【**생태관광의 촉진 등**】 정부는 동·식물의 서식지, 생태적으로 우수한 자연환경자산, 지역의 특색 있는 문화자산 등을 조화롭게 보존·복원 및 이용하여 이를 관광자원화하고 지역경제를 활성화함으로써 생태관광을 촉진하고, 국민 모두가 생태체험·교육의 장으로 활용할 수 있도록 하여야 한다.

第57条【**녹색성장을 위한 생산·소비 문화의 확산**】 ① 정부는 재화의 생산·소비·운반 및 폐기(이하 "생산 등"이라 한다)의 전 과정에서 에너지와 자원을 절약하고 효율적으로 이용하며 온실가스와 오염물질의 발생을 줄일 수 있도록 관련 시책을 수립·시행하여야 한다.

② 정부는 재화 및 서비스의 가격에 에너지 소비량 및 탄소배출량 등이 합리적으로 연계·반영되고 그 정보가 소비자에게 정확하게 공개·전달될 수 있도록 하여야 한다.
③ 정부는 재화의 생산 등의 전 과정에서 에너지와 자원의 사용량, 온실가스와 오염물질의 배출량 등을 분석·평가하고 그 결과에 관한 정보를 축적하여 이용할 수 있는 정보관리체계를 구축·운영할 수 있다.
④ 정부는 녹색제품의 사용·소비의 촉진 및 확산을 위하여 재화의 생산자와 판매자 등으로 하여금 그 재화의 생산등의 과정에서 발생되는 온실가스와 오염물질의 양에 대한 정보 또는 등급을 소비자가 쉽게 인식할 수 있도록 표시·공개하도록 하는 등의 시책을 수립·시행할 수 있다.

第58조【녹색생활 운동의 촉진】 ① 정부는 국민 및 기업들이 녹색생활에 친숙할 수 있도록 하는 시책을 마련하고 지방자치단체·기업·민간단체 및 기구 등과 협력체계를 구축하며 교육·홍보를 강화하는 등 범국민적 녹색생활 운동을 적극 전개하여야 한다.
② 정부는 녹색생활 운동이 민간주도형의 자발적 실천운동으로 전개될 수 있도록 관련 민간단체 및 기구 등에 대하여 필요한 재정적·행정적 지원 등을 할 수 있다.

第59조【녹색생활 실천의 교육·홍보】 ① 정부는 저탄소 녹색성장을 위한 교육·홍보를 확대함으로써 산업체와 국민 등이 저탄소 녹색성장을 위한 정책과 활동에 자발적으로 참여하고 일상생활에서 녹색생활 문화를 실천할 수 있도록 하여야 한다.
② 정부는 녹색생활 실천이 어릴 때부터 자연스럽게 이루어질 수 있도록 교과용 도서를 포함한 교재 개발 및 교원 연수 등 저탄소 녹색성장에 관한 학교교육을 강화하고 일반 교양교육, 직업교육, 기초평생교육 과정 등과 통합·연계한 교육을 강화

하여야 한다.
③ 정부는 녹색생활 문화의 정착과 확산을 촉진하기 위하여 신문·방송·인터넷포털 등 대중매체를 통한 교육·홍보 활동을 강화하여야 한다.
④ 공영방송은 지구온난화에 따른 기후변화 및 에너지 관련 프로그램을 제작·방영하고 공익광고를 활성화하도록 적극 노력하여야 한다.

제 7 장 보 칙

제60조【자료제출 등의 요구】 ① 위원회는 직무 수행상 필요하다고 인정되는 경우 관계 중앙행정기관·지방자치단체·공공기관의 장에게 저탄소 녹색성장에 관한 정보 또는 자료의 제출을 요구할 수 있다.
② 제1항에 따른 요구를 받은 관계 기관의 장은 국방상 또는 국가안전보장상 기밀을 요하는 사항 등 정당한 사유가 없으면 이에 응하여야 한다.

제61조【국제협력의 증진】 ① 정부는 외국 및 국제기구 등과 저탄소 녹색성장에 관한 정보교환, 기술협력 및 표준화, 공동조사·연구 등의 활동에 참여하여 국제협력, 국외진출의 증진을 도모하기 위한 각종 시책을 마련하도록 한다.
② 국가는 개발도상국가가 기후변화에 효과적으로 대응하고 지속가능발전을 촉진할 수 있도록 재정 지원을 하는 등 국제사회의 기대에 맞는 국가적 책무를 성실히 이행하고 국가의 외교적 위상을 높일 수 있도록 노력하여야 한다.
③ 정부는 국제기구 및 관련 기관에서 발표하는 공신력 있는

기후변화대응 평가에 대한 국가별 지수에서 우리나라의 위상 및 평가가 올라갈 수 있도록 기후변화대응을 적극 추진하고 국제협력을 강화하며 관련 정보를 충분히 제공하는 등 모든 노력을 기울여야 한다.

제62조【국회 보고】 ① 정부는 제9조 제1항에 따른 녹색성장 국가전략을 수립하였을 때에는 지체 없이 국회에 보고하여야 한다.
② 중앙행정기관의 장은 중앙추진계획을 수립하였을 때에는 지체 없이 소관 상임위원회(또는 관련 특별위원회)에 보고하여야 하며, 그 이행결과를 다음 해 2월 말일까지 소관 상임위원회(또는 관련 특별위원회)에 보고하여야 한다.

제63조【국가보고서의 작성】 ① 정부는 「기후변화에 관한 국제연합 기본협약」에서 정하는 바에 따라 국가보고서를 작성할 수 있다.
② 정부는 제1항에 따른 국가보고서를 작성하기 위하여 필요한 경우 관계 중앙행정기관의 장에게 자료의 제출을 요청할 수 있다. 이 경우 관계 중앙행정기관의 장은 특별한 사유가 없으면 요청에 따라야 한다.
③ 정부는 제1항에 따른 국가보고서를 「기후변화에 관한 국제연합 기본협약」의 당사국총회에 제출할 때에는 위원회의 심의를 거쳐야 한다.

제64조【과태료】 ① 다음 각 호의 자에게는 1천만원 이하의 과태료를 부과한다.

1. 제42조 제6항·제9항 또는 제44조 제1항에 따른 보고를 하지 아니하거나 거짓으로 보고한 자
2. 제42조 제8항에 따른 개선명령을 이행하지 아니한 자
3. 제42조 제9항에 따른 공개를 하지 아니한 자
4. 제44조 제2항에 따른 시정이나 보완 명령을 이행하지 아

니한 자

② 제1항에 따른 과태료는 대통령령으로 정하는 바에 따라 관계 행정기관의 장이 부과·징수한다.

부 칙 <제9931호, 2010. 1. 13.>

제1조【시행일】 이 법은 공포 후 3개월이 경과한 날부터 시행한다. 다만, 부칙 제4조 제12항 및 제13항에 따른 환경경영체제 인증의 녹색경영체제인증으로의 변경은 공포 후 1년 6개월이 경과한 날로부터 시행한다.

제2조【명세서 작성에 관한 특례】 관리업체는 제44조에도 불구하고 이 법 시행 첫해에는 과거 3년간의 온실가스 배출량 및 에너지 소비량에 대하여 명세서를 작성하여 보고하여야 한다.

제3조【녹색성장 국가전략에 관한 경과조치】 이 법 시행 당시 종전의 대통령 훈령 제239호에 따라 설치된 녹색성장위원회가 수립하여 국무회의의 심의를 거쳐 시행 중인 녹색성장 국가전략은 제9조에 따른 녹색성장국가전략으로 본다.

제4조【다른 법률의 개정】 [생략]

박 덕 영 (Deok-Young PARK)

연세대학교 법과대학 졸업
연세대학교 대학원 법학석사, 법학박사
국비유학시험 합격 (국제법 분야)
영국 University of Cambridge 법학석사 (LL.M.)
영국 University of Edinburgh 박사과정 마침
프로그램심의조정위원회 수석연구원 / 지적재산권팀장
숙명여자대학교 법과대학 조교수
사법시험, 외무고시, 행정고시, PSAT 출제위원
연세대학교 법학전문대학원 조교수 (현)
연세대학교 법학연구원 부원장 / EU법센터장 (현)
한국국제경제법학회 총무부회장 (현)

■ 주요 저서 ■

국제법 (공동), 박영사, 2010.3
미국법과 법률영어 (번역), 박영사, 2009
국제저작권과 통상문제 (공동), 세창출판사, 2009
국제법 기본조약집 / WTO 통상조약집, 박영사, 2007
국제경제법 (국제경제법학회 공동), 박영사, 2006

국제환경조약집

2010년 7월 15일 초판 인쇄
2010년 7월 20일 초판 발행

편 자 박 덕 영
발행인 이 방 원
발행처 세창출판사
서울 서대문구 냉천동 182 냉천빌딩 4층
전화 723-8660 팩스 720-4579
e-mail: sc1992@empal.com
http://www.sechangpub.co.kr
신고번호 제300-1990-63호

잘못 만들어진 책은 바꾸어 드립니다. **정가 22,000원**

ISBN 978-89-8411-312-1 93360